Man-Chung Han and Chu-Wan Kim

SECTIONAL HUMAN ANATOMY

Transverse, Sagittal and Coronal Sections
Correlated with
Computed Tomography and Magnetic Resonance Imaging

ILCHOKAK   Seoul

IGAKU-SHOIN   Tokyo · New York

AUTHORS' ADDRESS

Man-Chung Han, M.D.
Professor of Radiology
Department of Diagnostic Radiology,
College of Medicine,
Seoul National University
28, Yeongun-Dong, Chongro-Ku,
Seoul, Korea

Chu-Wan Kim, M.D.
Professor and Chairman
Department of Diagnostic Radiology,
College of Medicine,
Seoul National University
28, Yeongun-Dong, Chongro-Ku,
Seoul, Korea

ISBN 0-89640-118-9 (NEW YORK)
4-260-14118-X (TOKYO)

Printed and Bound in Korea

PREFACE

With the advent of new diagnostic modalities including ultrasonography, computed tomography and magnetic resonance imaging, there has been a steadily growing demand for atlases which display the human body in multidirectional section.

It is no longer only the anatomists who deal with pictures of sections of the human body on a daily basis. Today, clinicians in ever increasing numbers must do the same.

The purpose of this book is to provide a detailed depiction of sectional anatomy for medical students, residents, and practitioners of radiology, medicine and surgery.

The color photos of the anatomic specimens in this atlas have the advantage of being more realistic than previously used schematic drawings or black and white photos. Each anatomical structure is labeled directly on the photos for quick review, but not every structure is labeled in every illustration. Since each anatomic section is sequential, an unlabeled structure can be identified by its position in preceding or subsequent sections.

For easy understanding, segmental anatomy has been organized into three parts; head and neck, chest, and abdomen and pelvis. Each part consists of sequential anatomic sections and corresponding CT images, except in sagittal sections. In part four, magnetic resonance images are presented so that they can be compared with the anatomic and CT sections.

To make the information contained in this book accessible to a large audience the index terms are written in English, Chinese, and Korean.

We are deeply grateful to the many people whose assistance was invaluable in compiling this atlas. We would like to express our special thanks to the faculty of the Department of Anatomy for their assistance in preparing the cadavers. This atlas would not have been completed without help from the residents of our Department of Diagnostic Radiology, under the guidance of Dr. Jung-Gi Im, one of the contributors, who devoted his time and effort to this project.

We owe a special debt of gratitude to Professor Zang-Hee Cho, Korea Advanced Institute of Science and Technology, for his kind assistance in obtaining the magnetic resonance imaging.

We would also like to extend our thanks to Department of Radiology, Korea General Hospital, for the help in collecting CT images and to Mr. Young-Bo Shim for his excellent color photographs.

Finally, we would like to express our deep appreciation to our publisher, Ilchokak, whose understanding, patience and advice was of invaluable help.

September, 1985

Man-Chung Han, M.D.
Chu-Wan Kim, M.D.

머 리 말

최근 의학부분에 초음파촬영, 전산화단층촬영 및 자기공명영상촬영 등의 새로운 진단방법이 개발되어 임상에 이용됨에 따라, 인체의 단면해부학에 관한 지식은 해부학분야에서는 물론 일상 진료에 임하는 모든 임상의사에게도 필수적인 것이 되었다.

이러한 추세에 따라 외국에서는 단면해부학서적이 이미 발간되어 왔으나 인체단면이 흑백으로만 인쇄되었거나, 천연색상인 경우에도 상이 불분명하거나, 각 구조물을 표기하는데 있어서 간결하지 못한 점이 있어 아쉬움을 느끼고 있었다. 한편으로는 한국인을 위한 해부학서적은, 한국인 사체를 사용하여 제작되는 것이 바람직하다는 생각도 평소에 가지고 있었다. 이러한 이유로 하여 저자들은 3년 전부터 이 책의 발간준비를 시작하게 되었다.

이 책은 의과대학 학생을 비롯하여 전공의·내과계·외과계 및 방사선과학분야의 전문의를 위하여 제작되었으며, 인체의 단면과 해당부위의 전산화단층촬영영상을 대비시켰고 더 나아가서 일부분은 자기공명영상도 삽입하여 해부학 자체만이 아닌 임상진료에 이용될 수 있는 해부도보가 되도록 노력하였다.

인체의 65개의 각 단면을 모두 천연색상으로 인쇄하였으며, 단면에 나타난 각각의 구조물은 독자로 하여금 쉽게 파악할 수 있도록 단면사진에 직접 약어로 된 문자로 표기하였다. 모든 구조물을 표기하려고 노력하였으나, 앞뒤로 중복되는 구조물인 경우는 생략된 경우도 있다. 한편 색인은 영문, 한문 및 한글을 차례로 기술하여 해부학적 구조물 표기에 일관성이 있도록 하였다.

우리나라 의학교육 및 의료계의 현실에 따라 부득이 영문학명 위주로 표기하였으며 이에 따라 이 책이 외국인에게도 널리 애용될 수 있으리라 믿는다.

이 책에 게재된 자기공명영상은 한국과학기술원에서 개발한 0.15 Tesla의 상온도자석 자기공명영상장치로 촬영한 것으로 그 의의가 크다고 할 수 있다.

이 책에 사용된 사체의 고정 및 보관과 색인정리에 적극적인 협력과 지도를 아끼지 않았던 서울대학교 의과대학 해부학교실 교수진에게 감사를 드리고 자기공명영상을 수록할 수 있도록 협조하여 주신 한국과학기술원의 조장희교수 및 대학원생 여러분과 전산화단층촬영사진 수집에 협력해 주신 고려병원 방사선과에 감사의 말씀을 드린다.

선명한 천연색 및 흑백사진을 얻기 위하여 열과 성을 다하였던 서울대학교병원 사진실 심영보씨의 노고를 치하하고 아울러 본 저서의 출간이 가능하도록 헌신적인 노력을 하여 준 서울대학교 의과대학 진단방사선과학교실 교수 전원 및 의국원들에게 감사하며 특히 사체절단에서부터 출판과정의 세부사항까지 일선에서 일을 관장한 임정기교수의 노고는 특기하여야만 할 것이다. 끝으로 이해와 인내를 가지고 꾸준히 협조를 아끼지 않았던 일조각 여러분에게 감사한다.

1985년 9월 일

한 만 청

김 주 완

INTRODUCTION

An excellent way to understand the three dimensional anatomical structure of the human body is by studying multidirectional sections of a cadaver with corresponding radiographic images.

The ultimate objective of this atlas is to portray normal sectional anatomy and anatomical relationships in basic three dimensional form.

A. Preparation of the Cadaver

Immediately after the cadaver is brought to the Department of Anatomy it is embalmed by injecting a preserving fluid, consisting of 10% formalin and glycerine, through one of the femoral arteries. The cadaver is then placed in the embalming machine for one month, which facilitates fixation of the tissue with the aid of high atmospheric pressure.

The major arterial and venous channels are filled with red and blue dye, respectively, for the sake of artificial accentuation of the vascular tree. Dye is injected through the femoral artery and vein with an application of 300mmHg and 200mmHg pressure. Some of the dye extravasates through unidentified routes to the extravascular potential spaces of the pleura, pericardium and peritoneum. Though this is an undesirable side effect, it demonstrates clearly the potential spaces or fascial planes. The mixture of gelatin solution and water soluble dye provides excellent solidification of the dye in the vascular space, thus preventing leakage and intrusion to adjacent structures during saw cutting of the cadaver.

After the period of fixation and dye infusion, the cadaver is placed in horizontal position on the cabinet of the freezing room. With an additional application of dry ice around the cadaver, the body is maintained in anatomical position for the complete freezing period of 48 hours.

Complete freezing of the body provides uniform hardness of the tissue so as to present uniform resistence to the progression of the band saw.

B. Section and Preparation of Anatomical Slices

For ease of handling, extremities are removed. The head is cut into slices with the reference plane of orbitomeatal line in transverse section. The adjustable guiding plate parallel to the saw blade is tightly attached to the surface of each slice in order to obtain uniform thickness.

The planes of reference in the transverse section of the chest, abdomen and pelvis are both nipples and the anterior superior iliac spine. The thickness of each slice varies according to the location, ranging from 1cm to 1.5cm.

Immediately after sectioning, each slice is placed on a glass plate and the surface of the slice is washed with running water and a gauze. Then, another glass plate is placed on the exposed side and the slice is turned over.

The cleaned slices are covered with cloth and glass plate to prevent excessive wetness and dryness of the surface. They are then stored in the refrigerating room for not more than 48 hours.

Before photographing, some of the extravasated dye and contents of the natural cavities are removed.

C. Photography

The photographic equipment consists of a Mamiya RB 67 Camera with Sekkor 90mm 3.8 lens. Exposure time is 1/60 sec with the aperture set at 8. Two 500W lamps are placed 1.5m away from the specimen at a 45 degree angle. Kodak CII film (70×60mm) is used.

Pictures are taken of both sides of each slice. Photographs of the transverse section of the head and neck are enlarged 1.5 times while sagittal sections of the body are reduced to 0.75 times their natural size.

The photographs have been chosen and arranged to a large extent on the basis of their clinical importance. So, transverse sections are viewed from the bottom side. The sagittal section of the chest is arranged with the anterior chest at the reader's left side because of familiarity.

The abdominal and pelvic sagittal sections are arranged in reversed order for the convenience of ultrasonogram interpretation.

CT scan pictures were taken with General Electric CT/T 9800 or 8800 scanner. Though we tried to select identical pictures for comparison with corresponding anatomic slices this was not possible in all cases because the location of internal organs varies.

Magnetic resonance images are from the resistive magnet system machine operating at 0.15 Tesla. All the images were taken using the spin-echo technique at TR/TE 400msec/30msec.

D. Labelling

A great effort has been made to label the structures in as much as possible. Abbreviations placed directly on the photograph allow for prompt recognition. For a single word, the abbreviation consists of the first two or three letters, and for multiple words, the first letter of each word is used in capital letters. For the sake of clarity, each abbreviation corresponds to only one structure.

Jung-Gi Im, M.D.

CONTENTS

CONTRIBUTORS

From   Department of Diagnostic Radiology,
        College of Medicine,
        Seoul National University,
        Seoul, Korea.

Kee-Hyun Chang, M.D.
   Assistant Professor

Byung-Ihn Choi, M.D.
   Assistant Professor

Man-Chung Han, M.D.
   Professor

Jung-Gi Im, M.D.
   Instructor

Heung-Sik Kang, M.D.
   Instructor

Chu-Wan Kim, M.D.
   Professor and Chairman

Jae-Hyung Park, M.D.
   Assistant Professor

Kyung-Mo Yeon, M.D.
   Assistant Professor

PART 1
BRAIN, HEAD and NECK

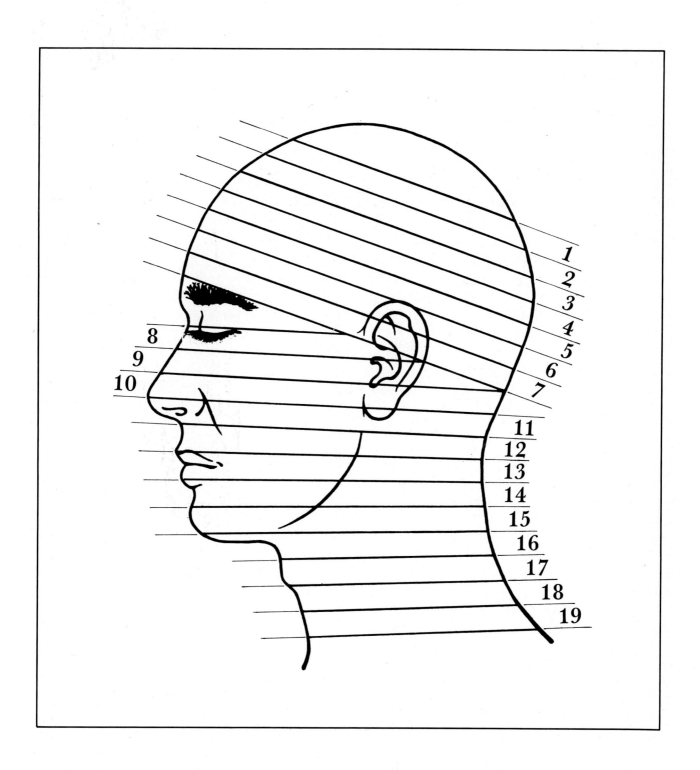

Transverse Section (Specimen & CT) Plate 1-19

Plate 1. TRANSVERSE BRAIN

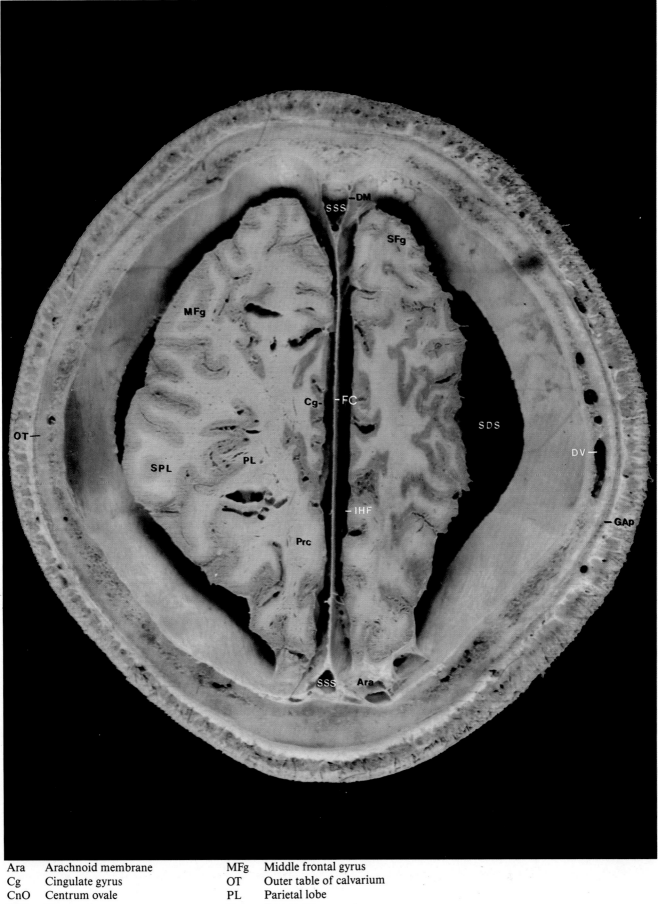

Ara	Arachnoid membrane	MFg	Middle frontal gyrus
Cg	Cingulate gyrus	OT	Outer table of calvarium
CnO	Centrum ovale	PL	Parietal lobe
DM	Dura mater	Prc	Precuneus
DV	Diploic vein	SDS	Subdural space
FC	Falx cerebri	SFg	Superior frontal gyrus
FL	Frontal lobe	SPL	Superior parietal lobule
GAp	Galea aponeurotica	SSS	Superior sagittal sinus
IHF	Interhemispheric flssure		

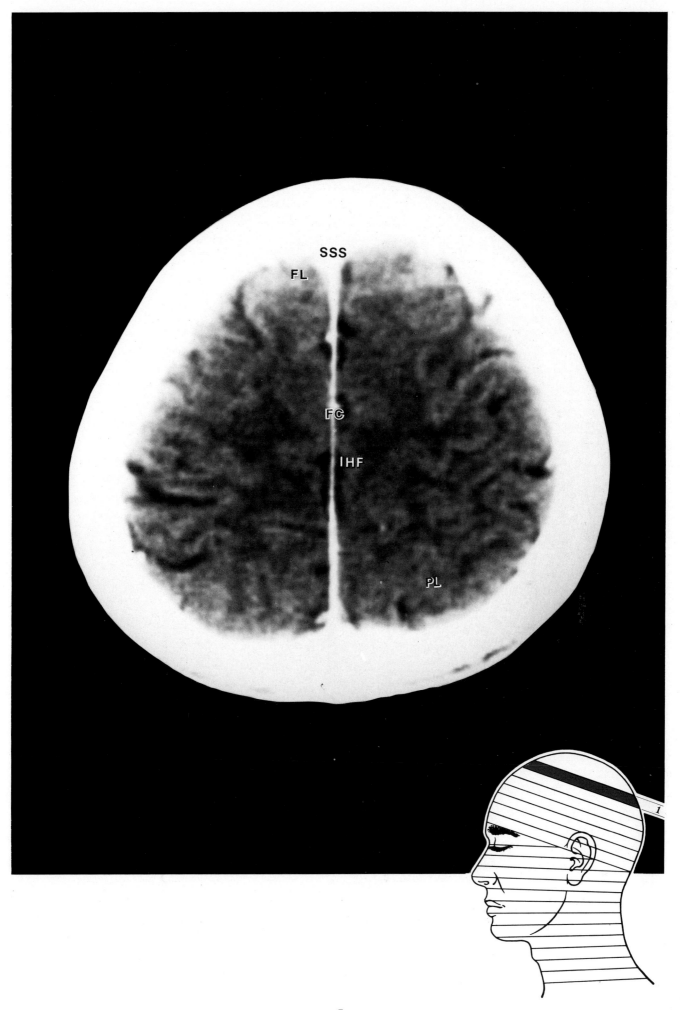

Plate 2. TRANSVERSE BRAIN

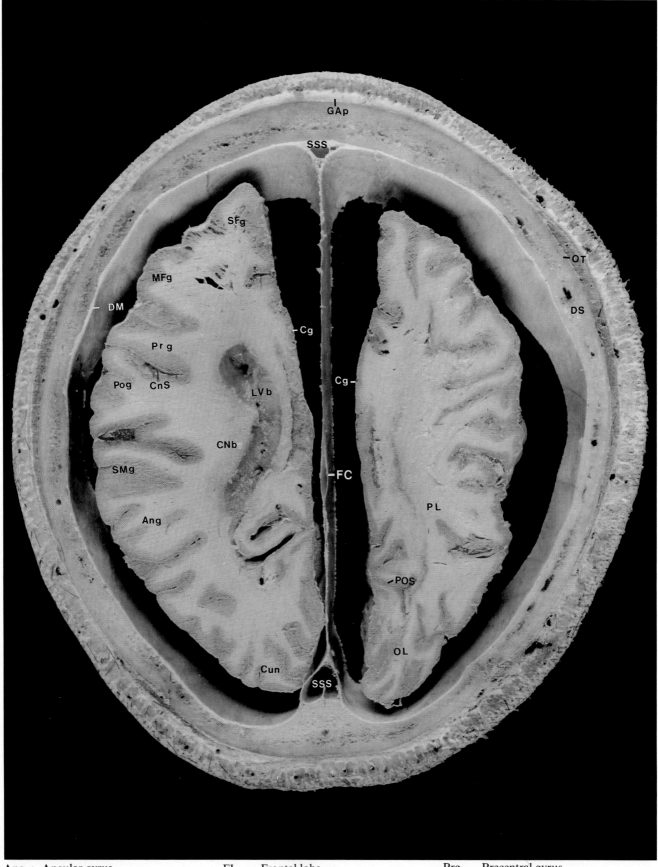

Ang	Angular gyrus	FL	Frontal lobe	Prg	Precentral gyrus
CNb	Caudate nucleus, body	GAp	Galea aponeurotica	SFg	Superior frontal gyrus
Cg	Cingulate gyrus	LVb	Lateral ventricle, body	SMg	Supramarginal gyrus
CnO	Centrum ovale	MFg	Middle frontal gyrus	SSS	Superior sagittal sinus
CnS	Central sulcus	OL	Occipital lobe		
Cun	Cuneus	OT	Outer table of calvarium		
DM	Dura mater	PL	Parietal lobe		
DS	Diploic space	POS	Parieto-occipital sulcus		
FC	Falx cerebri	Pog	Postcentral gyrus		

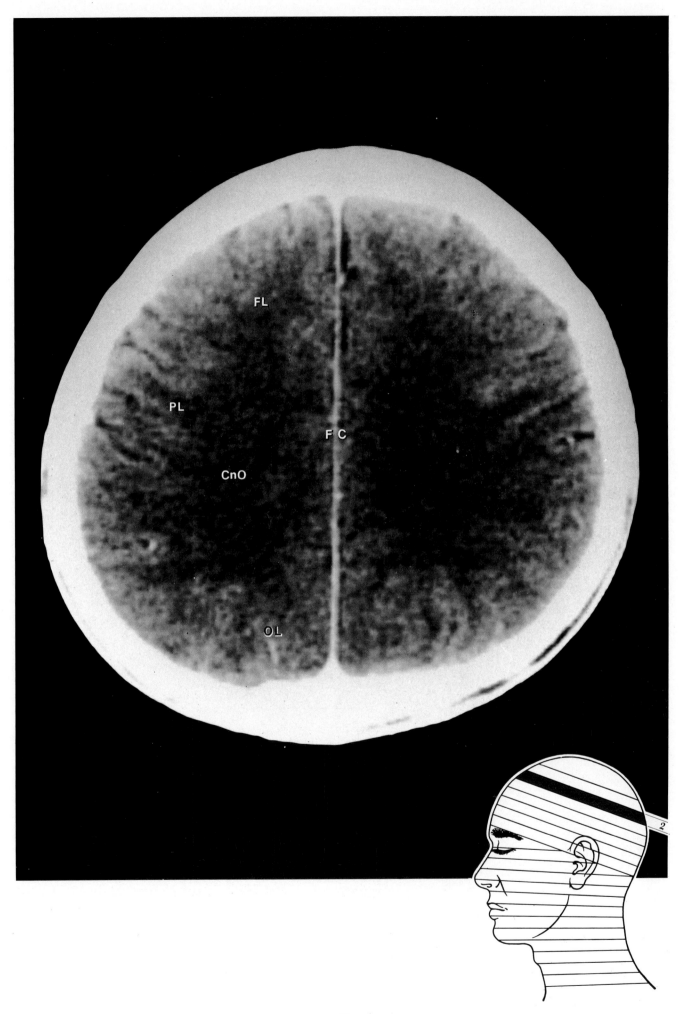

Plate 3. TRANSVERSE BRAIN

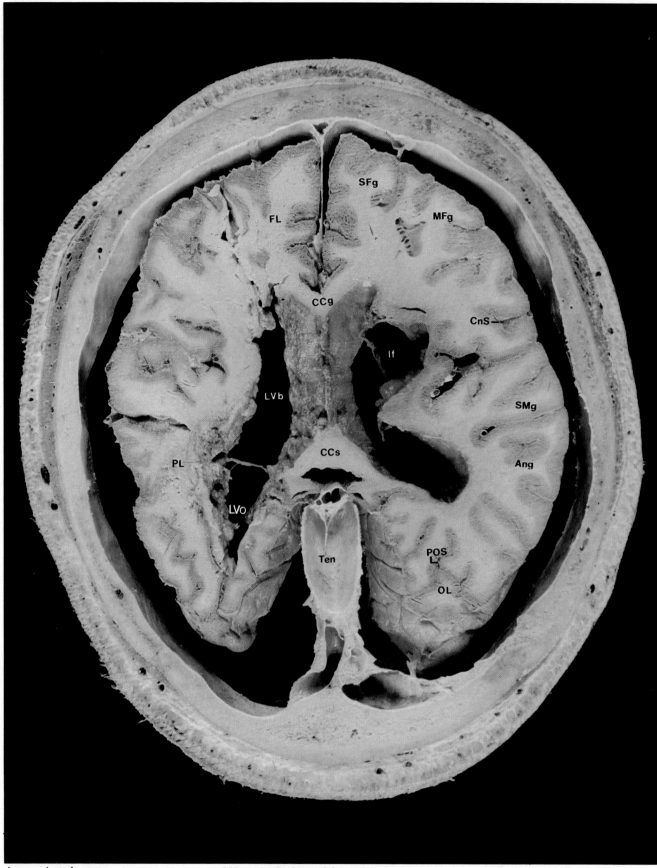

Ang	Angular gyrus	LVo	Lateral ventricle, occipital horn
CCg	Corpus callosum, genu	MFg	Middle frontal gyrus
CCs	Corpus callosum, splenium	OL	Occipital lobe
CR	Corona radiata	PL	Parietal lobe
CnS	Central sulcus	POS	Parieto-occipital sulcus
FC	Falx cerebri	SFg	Superior frontal gyrus
FL	Frontal lobe	SMg	Supramarginal gyrus
If	Old infarct	Ten	Tentorium cerebelli
LVb	Lateral ventricle, body		

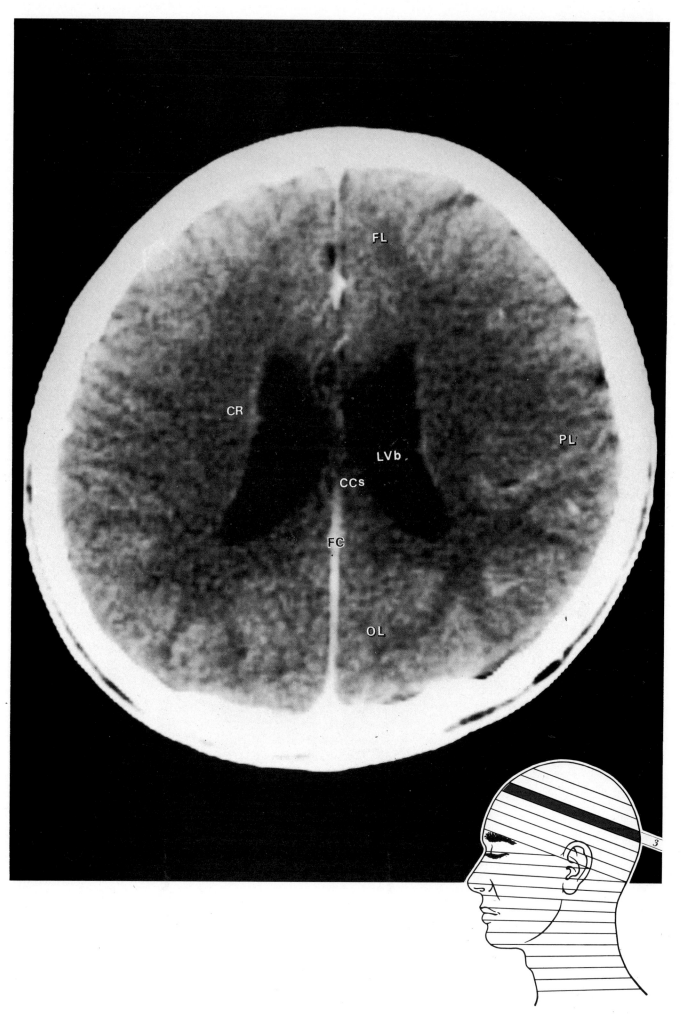

Plate 4. TRANSVERSE BRAIN

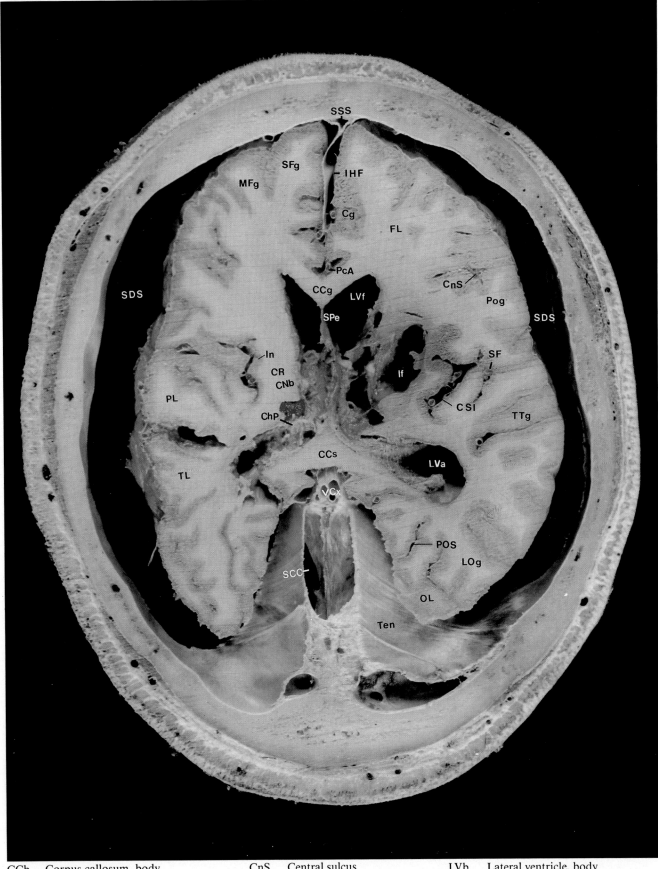

CCb	Corpus callosum, body	CnS	Central sulcus	LVb	Lateral ventricle, body
CCg	Corpus callosum, genu	FC	Falx cerebri	LVf	Lateral ventricle, frontal horn
CCs	Corpus callosum, splenium	FL	Frontal lobe	LVo	Lateral ventricle, occipital horn
CNb	Caudate nucleus, body	ICV	Internal cerebral vein	MFg	Middle frontal gyrus
CNh	Caudate nucleus, head	IHF	Interhemispheric fissure	OL	Occipital lobe
CR	Corona radiata	If	Old infarct	PL	Parietal lobe
CSI	Circular sulcus of insula	In	Insula	POS	Parieto-occipital sulcus
Cg	Cingulate gyrus	LOg	Lateral occipital gyrus	PcA	Pericallosal artery
ChP	Choroid plexus of lateral ventricle	LVa	Lateral ventricle, antrum	Pog	Postcentral gyrus

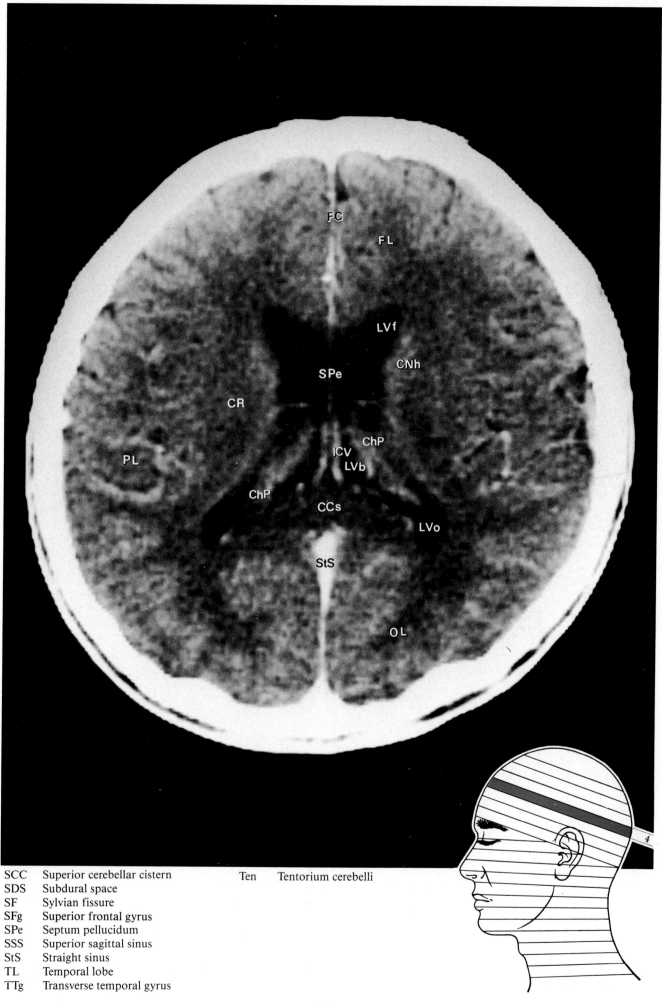

SCC	Superior cerebellar cistern
SDS	Subdural space
SF	Sylvian fissure
SFg	Superior frontal gyrus
SPe	Septum pellucidum
SSS	Superior sagittal sinus
StS	Straight sinus
TL	Temporal lobe
TTg	Transverse temporal gyrus

Ten   Tentorium cerebelli

Plate 5. TRANSVERSE BRAIN

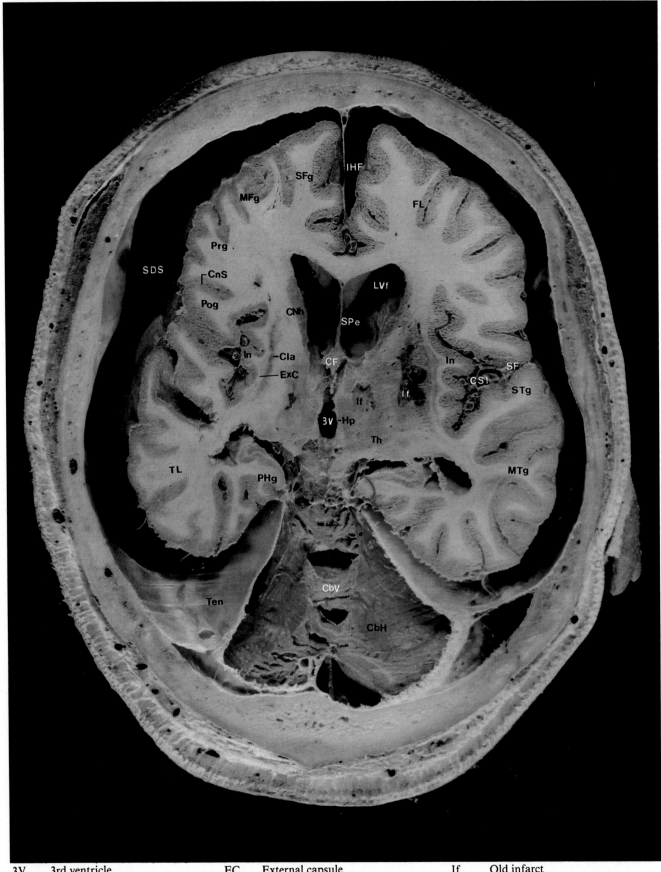

3V	3rd ventricle	**EC**	External capsule	**If**	Old infarct
CCb	Corpus callosum, body	**ExC**	Extreme capsule	**In**	Insula
CF	Column of fornix	**FC**	Falx cerebri	**LVf**	Lateral ventricle, frontal horn
CNh	Caudate nucleus, head	**FL**	Frontal lobe	**MFg**	Middle frontal gyrus
CSl	Circular sulcus of insula	**Hp**	Hypothalamus	**MTg**	Middle temporal gyrus
CbH	Cerebellar hemisphere	**ICV**	Internal cerebral vein	**OL**	Occipital lobe
CbV	Cerebellar vermis	**ICa**	Internal capsule, anterior limb	**PHg**	Parahippocampal gyrus
Cla	Claustrum	**ICp**	Internal capsule, posterior limb	**Pog**	Postcentral gyrus
CnS	Central sulcus	**IHF**	Interhemispheric fissure	**Prg**	Precentral gyrus

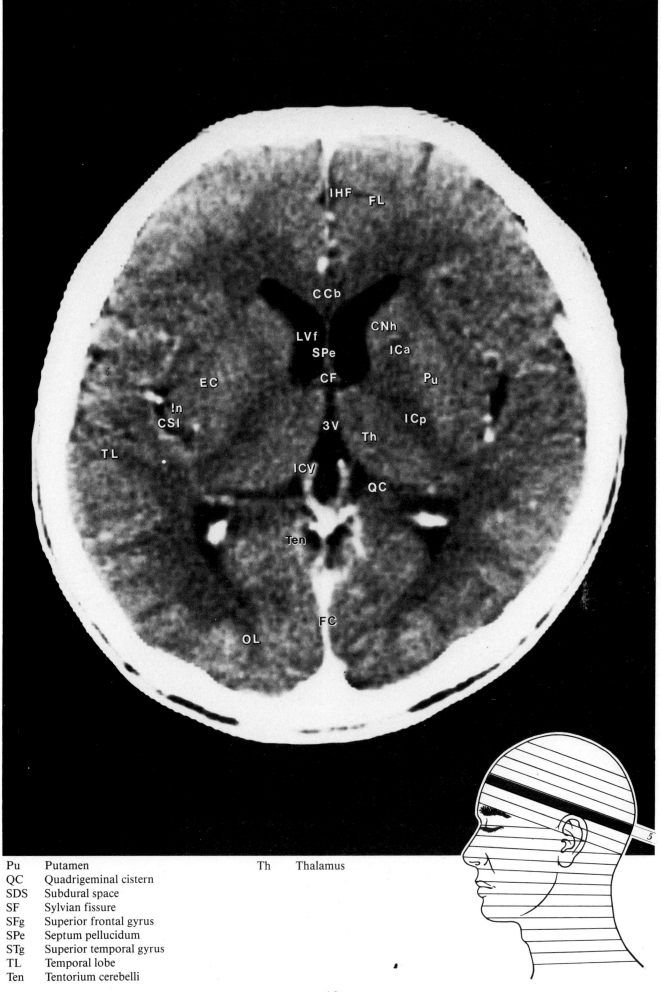

Pu	Putamen		Th	Thalamus
QC	Quadrigeminal cistern			
SDS	Subdural space			
SF	Sylvian fissure			
SFg	Superior frontal gyrus			
SPe	Septum pellucidum			
STg	Superior temporal gyrus			
TL	Temporal lobe			
Ten	Tentorium cerebelli			

Plate 6. TRANSVERSE BRAIN

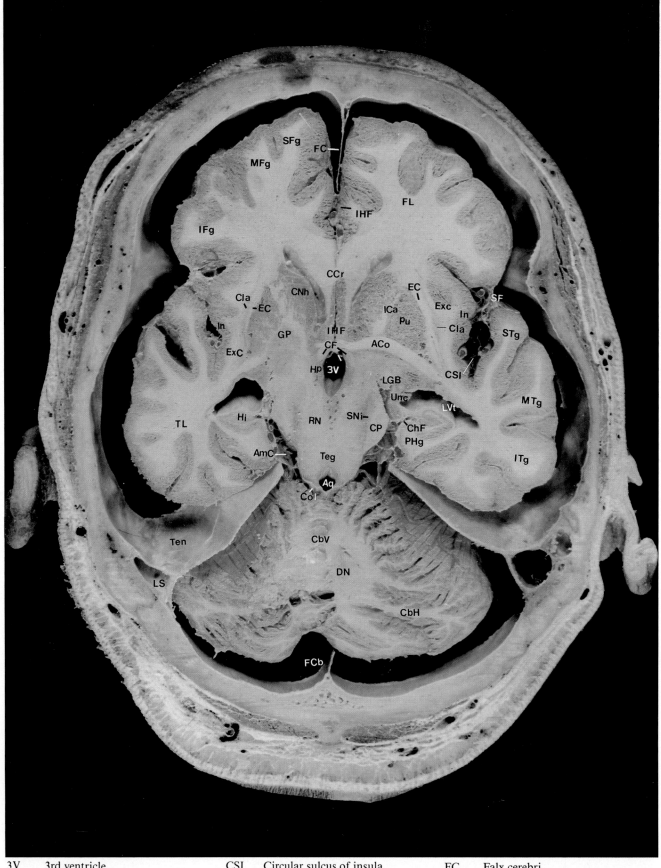

3V	3rd ventricle	CSI	Circular sulcus of insula	FC	Falx cerebri
ACo	Anterior commissure	CbH	Cerebellar hemisphere	FCb	Falx cerebelli
AmC	Ambient cistern	CbV	Cerebellar vermis	FL	Frontal lobe
Aq	Cerebral aqueduct	ChF	Choroidal fissure	GP	Globus pallidus
CCg	Corpus callosum, genu	Cla	Claustrum	Hi	Hippocampus
CCr	Corpus callosum, rostrum	Col	Colliculi (Quadrigeminal plate)	Hp	Hypothalamus
CF	Column of fornix	DN	Dentate nucleus	ICa	Internal capsule, anterior limb
CNh	Caudate nucleus, head	EC	External capsule	IFg	Inferior frontal gyrus
CP	Cerebral peduncle	ExC	Exterme capsule	IHF	Interhemispheric fissure

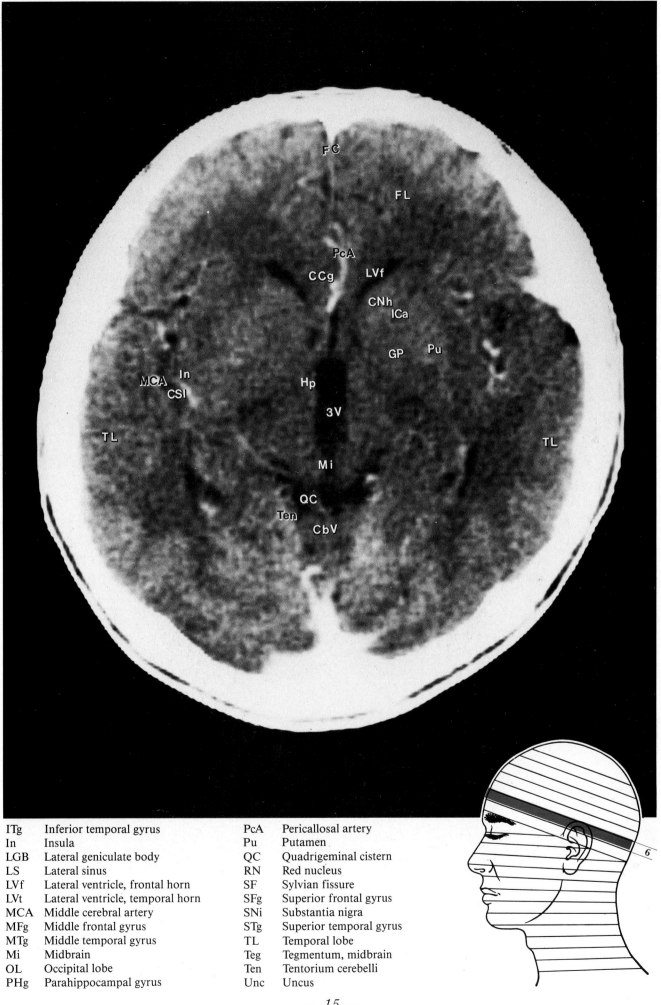

ITg	Inferior temporal gyrus	PcA	Pericallosal artery
In	Insula	Pu	Putamen
LGB	Lateral geniculate body	QC	Quadrigeminal cistern
LS	Lateral sinus	RN	Red nucleus
LVf	Lateral ventricle, frontal horn	SF	Sylvian fissure
LVt	Lateral ventricle, temporal horn	SFg	Superior frontal gyrus
MCA	Middle cerebral artery	SNi	Substantia nigra
MFg	Middle frontal gyrus	STg	Superior temporal gyrus
MTg	Middle temporal gyrus	TL	Temporal lobe
Mi	Midbrain	Teg	Tegmentum, midbrain
OL	Occipital lobe	Ten	Tentorium cerebelli
PHg	Parahippocampal gyrus	Unc	Uncus

Plate 7. TRANSVERSE BRAIN

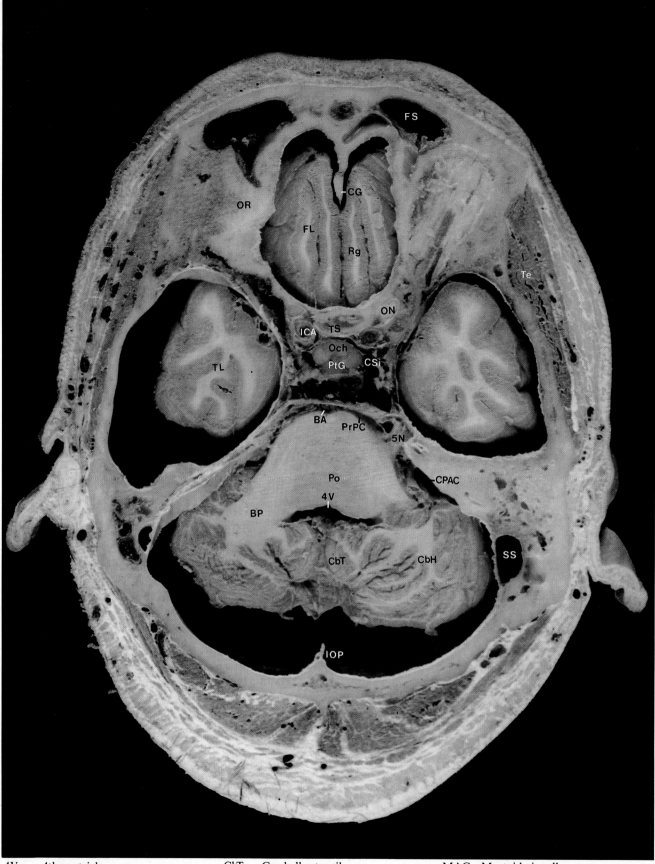

4V	4th ventricle	CbT	Cerebellar tonsil	MAC	Mastoid air cells
5N	5th cranial nerve	CbV	Cerebellar vermis	MCA	Middle cerebral artery
ACA	Anterior cerebral artery	FC	Falx cerebri	OCh	Optic chiasm
BA	Basilar artery	FL	Frontal lobe	ON	Optic nerve
BP	Brachium pontis	FS	Frontal sinus	OR	Orbital roof
CG	Crista galli	ICA	Internal carotid artery	PCA	Posterior cerebral artery
CPAC	Cerebellopontine angle cistern	IHF	Interhemispheric fissure	Po	Pons
CSi	Cavernous sinus	IOP	Internal occipital protuberance	PrPC	Pre-pontine cistern
CbH	Cerebellar hemisphere	LVt	Lateral ventricle, temporal horn	PtG	Pituitary gland

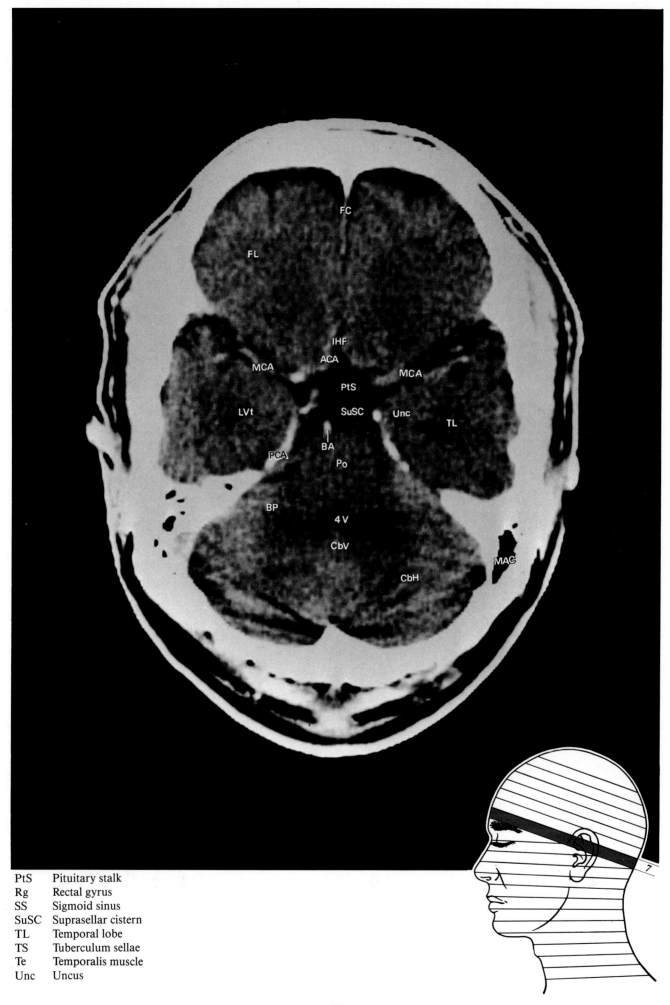

PtS Pituitary stalk
Rg Rectal gyrus
SS Sigmoid sinus
SuSC Suprasellar cistern
TL Temporal lobe
TS Tuberculum sellae
Te Temporalis muscle
Unc Uncus

Plate 8. TRANSVERSE HEAD AND NECK

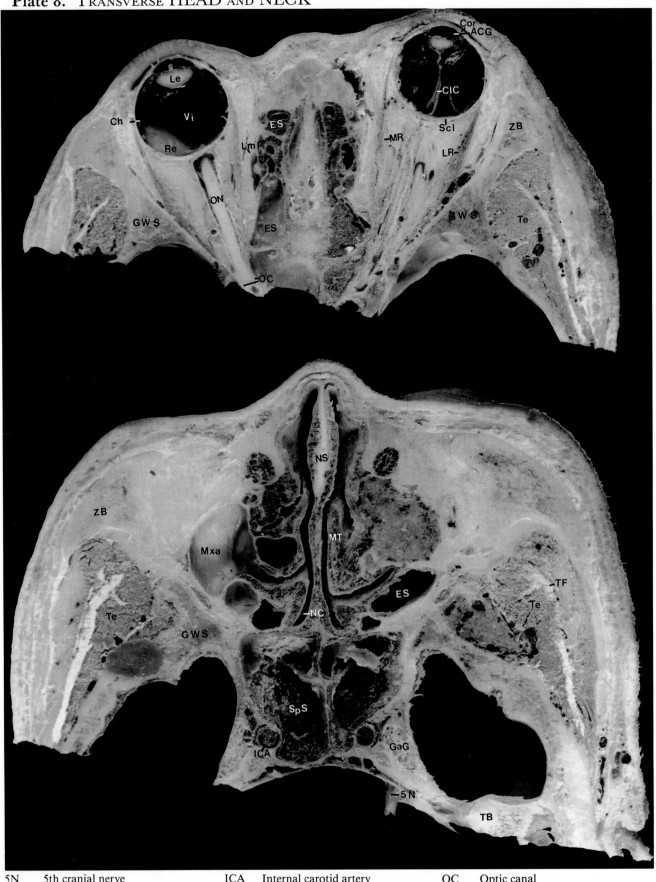

5N	5th cranial nerve	ICA	Internal carotid artery	OC	Optic canal
ACG	Anterior chamber of globe	LR	Lateral rectus muscle	ON	Optic nerve
ClC	Cloquet canal	Le	Lens	Re	Retina
Ch	Choroid	LmP	Lamina papiracea	Scl	Sclera
Cor	Cornea	MR	Medial rectus muscle	SpS	Sphenoid sinus
ES	Ethmoid sinus	MT	Middle turbinate	TB	Temporal bone
Ey	Eyeball	Mxa	Maxillary sinus, antrum	TF	Temporalis fascia
GWS	Greater wing of sphenoid	NC	Nasal cavity	TL	Temporal lobe
GaG	Gasselian ganglion	NS	Nasal septum	Te	Temporalis muscle

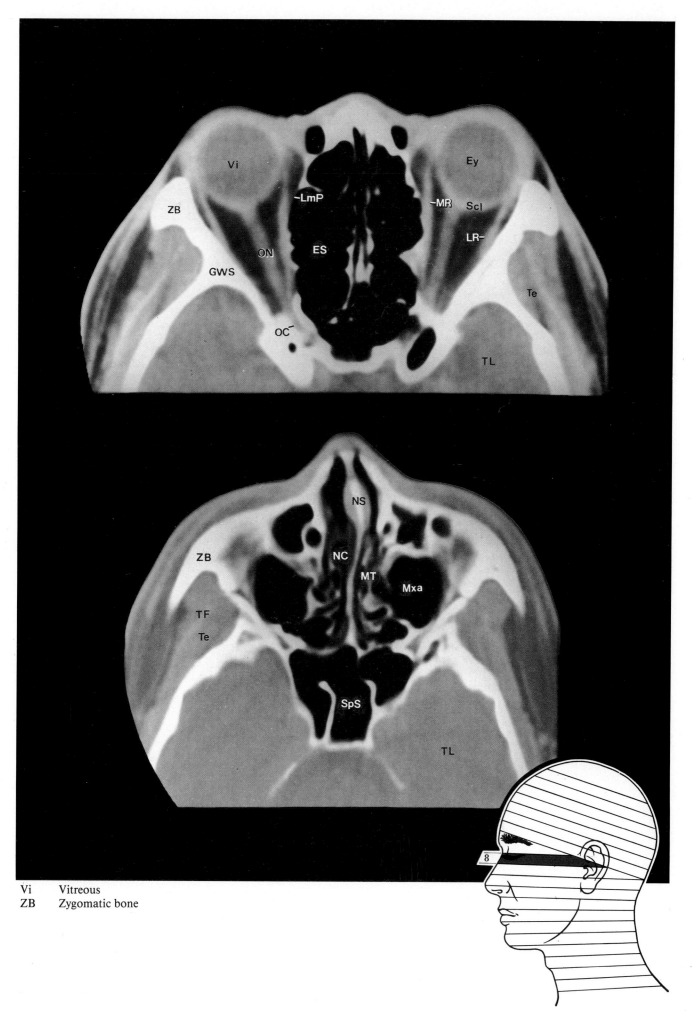

Vi   Vitreous
ZB   Zygomatic bone

Plate 9. TRANSVERSE HEAD AND NECK

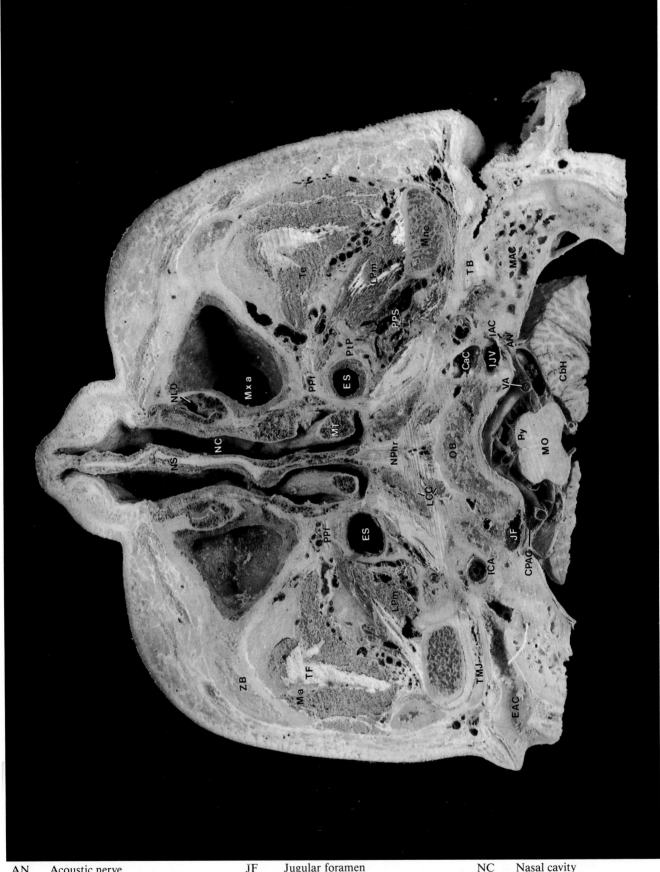

AN	Acoustic nerve	JF	Jugular foramen	NC	Nasal cavity
CPAC	Cerebellopontine angle cistern	LCC	Longus capitis & colli muscle	NLD	Nasolacrimal duct
CaC	Carotid canal	LPm	Lateral pterygoid muscle	NS	Nasal septum
CbH	Cerebellar hemisphere	MAC	Mastoid air cells	NPh	Nasopharynx
EAC	External auditory canal	MO	Medulla oblongata	NPhr	Nasopharynx, roof
ES	Ethmoid sinus	MT	Middle turbinate	OB	Occipital bone
IAC	Internal auditory canal	Ma	Masseter muscle	PPS	Parapharyngeal space
ICA	Internal carotid artery	Mnc	Mandible, condylar process	PPf	Pterygopalatine fossa
IJV	Internal jugular vein	Mxa	Maxillary sinus, antrum	PtP	Pterygoid plate

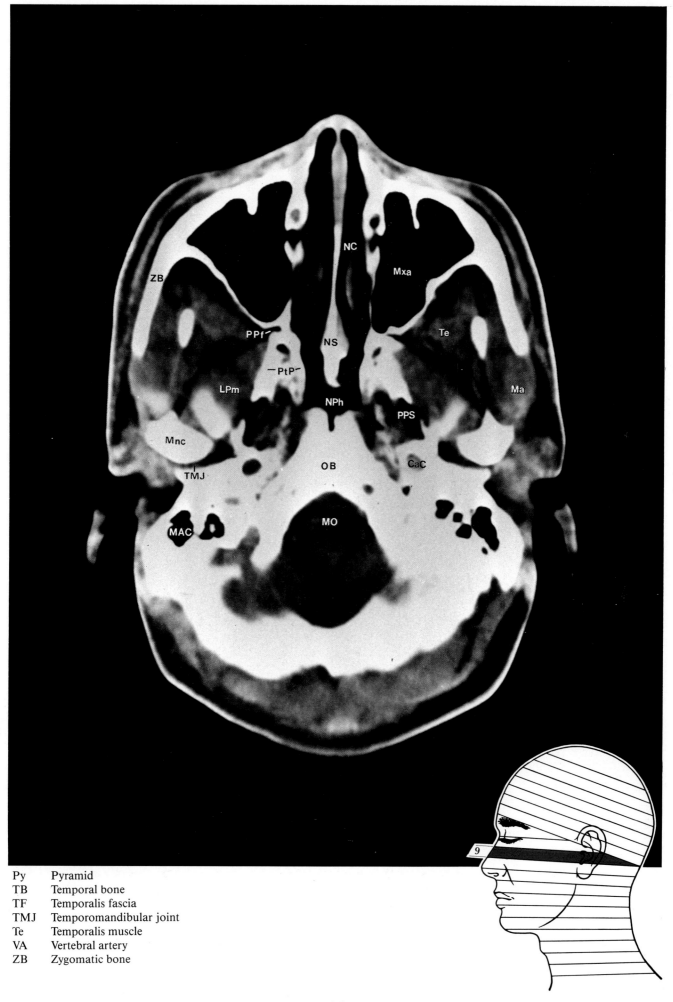

Py   Pyramid
TB   Temporal bone
TF   Temporalis fascia
TMJ Temporomandibular joint
Te   Temporalis muscle
VA   Vertebral artery
ZB   Zygomatic bone

Plate 10. TRANSVERSE HEAD AND NECK

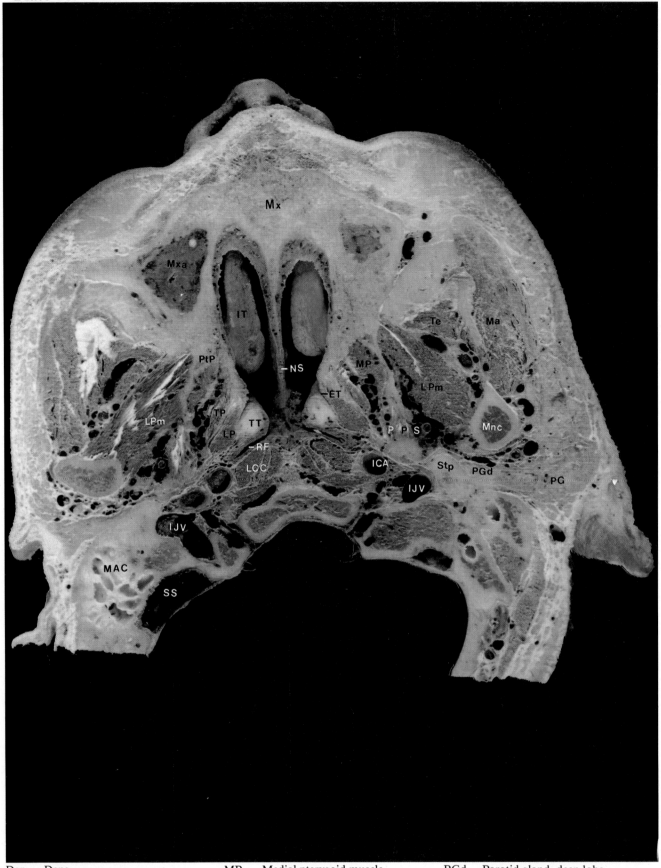

Den	Dens	MP	Medial pterygoid muscle	PGd	Parotid gland, deep lobe
ET	Eustachian tube	Ma	Masseter muscle	PPS	Parapharyngeal space
ICA	Internal carotid artery	Mnc	Mandible, condylar process	PtP	Pterygoid plate
IJV	Internal jugular vein	Mnr	Mandible, ramus	RF	Rosenmuller fossa
IT	Inferior turbinate	Mx	Maxilla	SS	Sigmoid sinus
LCC	Longus capitis & colli muscle	Mxa	Maxillary sinus, antrum	Stp	Styloid process
LP	Levator palati muscle	NS	Nasal septum	TP	Tensor palati muscle
LPm	Lateral pterygoid muscle	NPh	Nasopharynx	TT	Torus tubarius
MAC	Mastoid air cells	PG	Parotid gland	Te	Temporalis muscle

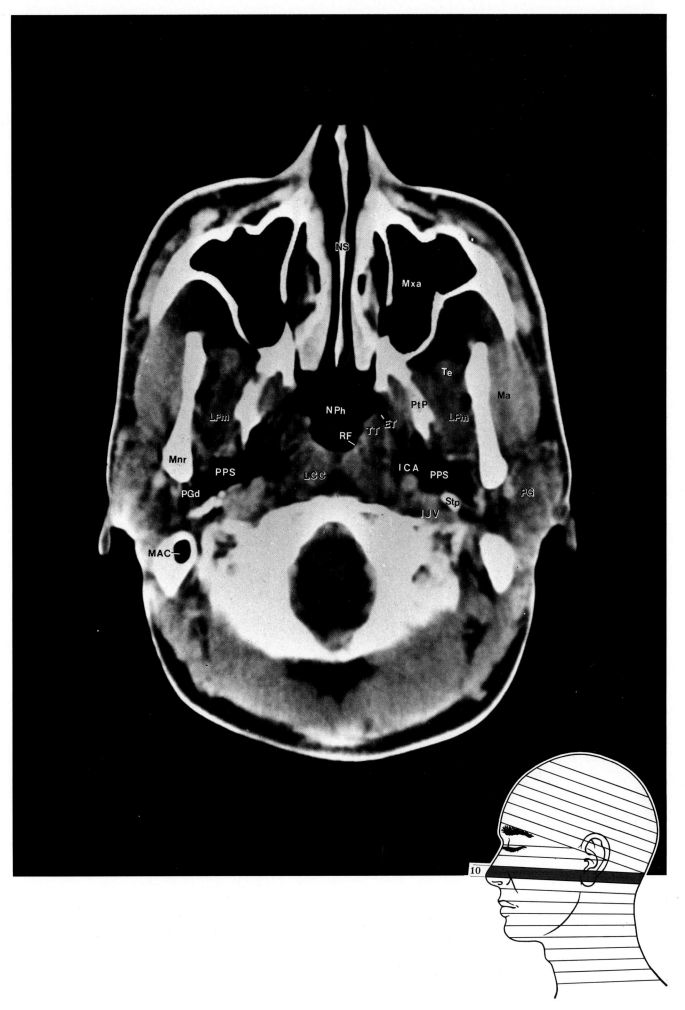

Plate 11. TRANSVERSE HEAD AND NECK

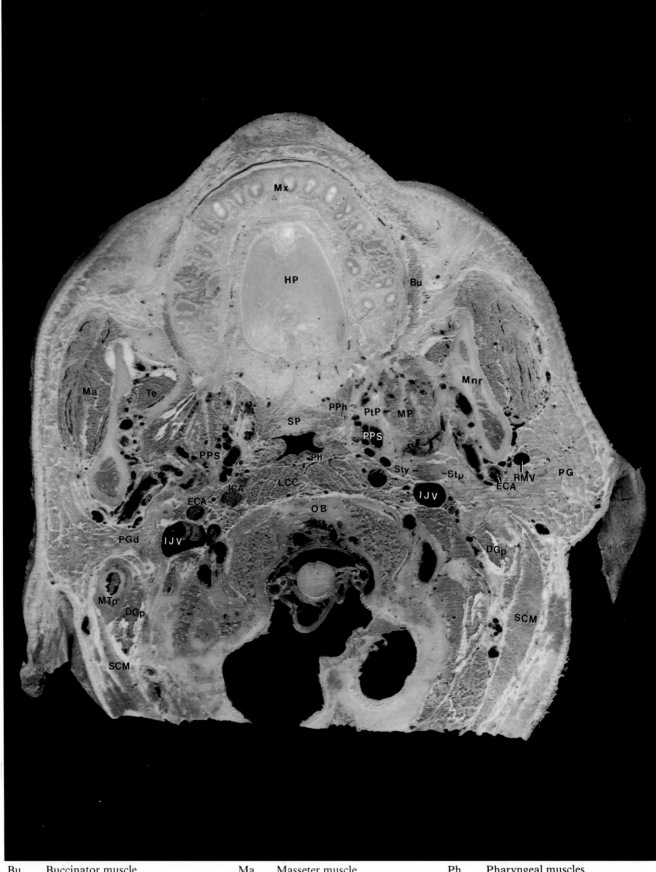

Bu	Buccinator muscle		Ma	Masseter muscle		Ph	Pharyngeal muscles
DGp	Digastric muscle, posterior belly		Mnr	Mandible, ramus		PtP	Pterygoid plate
ECA	External carotid artery		Mx	Maxilla		RMV	Retromandibular vein
HP	Hard palate		NPh	Nasopharynx		SCM	Sternocleidomastoid muscle
ICA	Internal carotid artery		OB	Occipital bone		SP	Soft palate
IJV	Internal jugular vein		PG	Parotid gland		Stp	Styloid process
LCC	Longus capitis & colli muscle		PGd	Parotid gland, deep lobe		Sty	Styloid muscle
MP	Medial pterygoid muscle		PPS	Parapharyngeal space		Te	Temporalis muscle
MTp	Mastoid tip		PPh	Palatopharyngeus muscle			

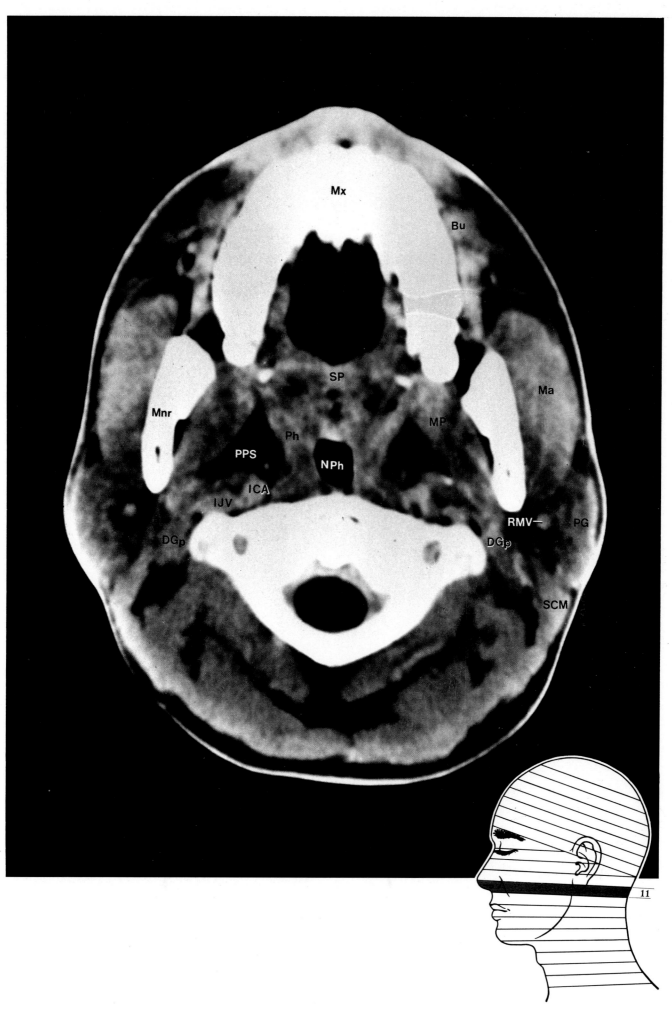

Plate 12. Transverse HEAD AND NECK

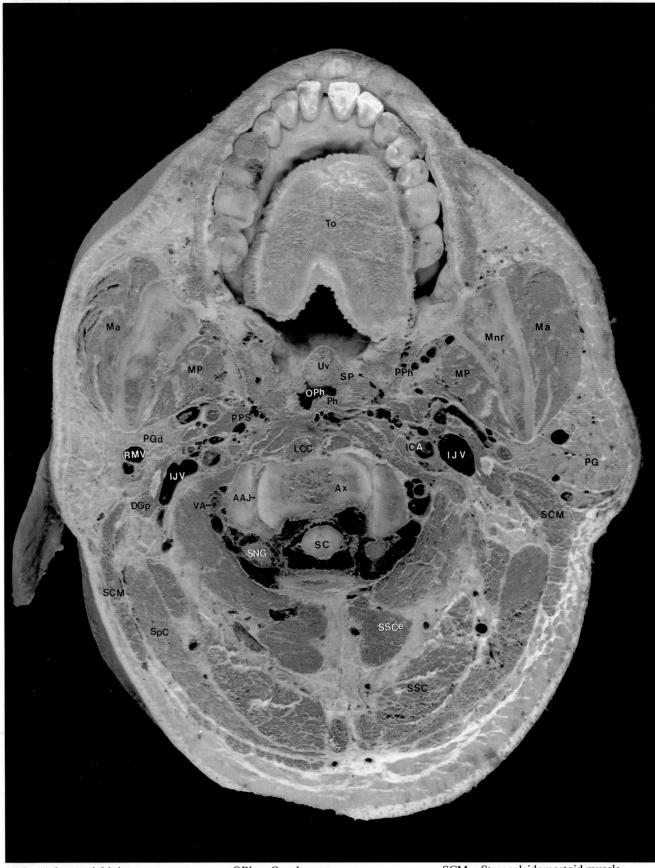

AAJ	Atlantoaxial joint	OPh	Oropharynx	SCM	Sternocleidomastoid muscle
Ax	Axis	PG	Parotid gland	SNG	Spinal nerve ganglion
DGp	Digastric muscle, postrior belly	PGd	Parotid gland, deep lobe	SP	Soft palate
ICA	Internal carotid artery	PPS	Parapharyngeal space	SSC	Semispinalis capitis muscle
IJV	Internal jugular vein	PPh	Palatopharyngeus muscle	SSCe	Semispinalis cervicalis muscle
LCC	Longus capitis & colli muscle	Ph	Pharyngeal muscle	SpC	Splenius capitis muscle
MP	Medial pterygoid muscle	RMV	Retromandibular vein	Stp	Styloid process
Ma	Masseter muscle	SAS	Subarachnoid space	To	Tongue
Mnr	Mandible, ramus	SC	Spinal cord	Uv	Uvula

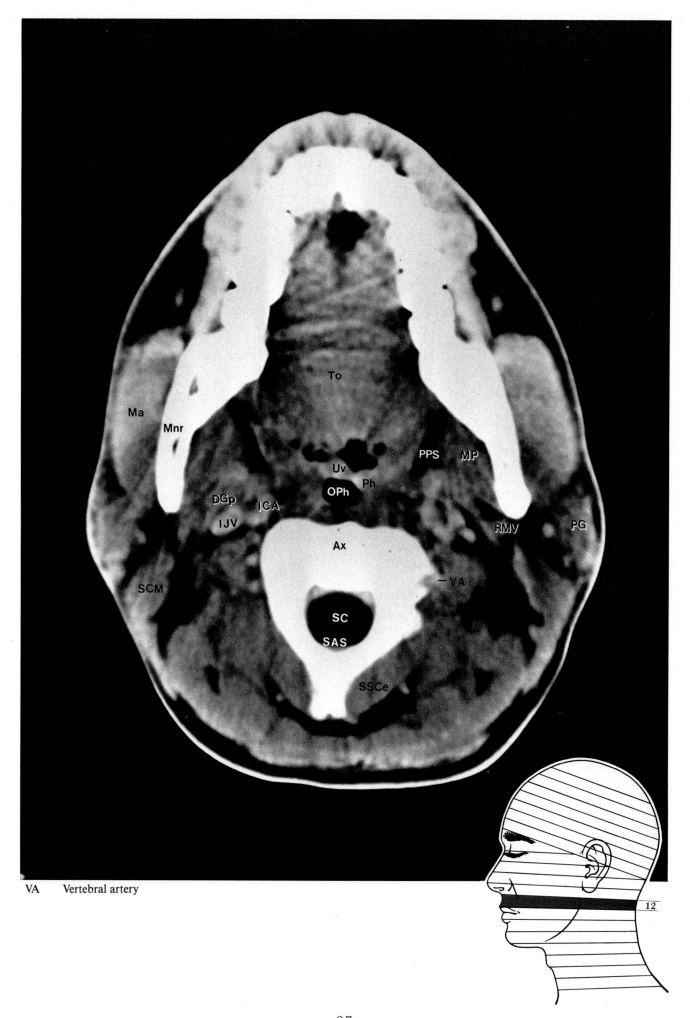

VA   Vertebral artery

Plate 13. TRANSVERSE HEAD AND NECK

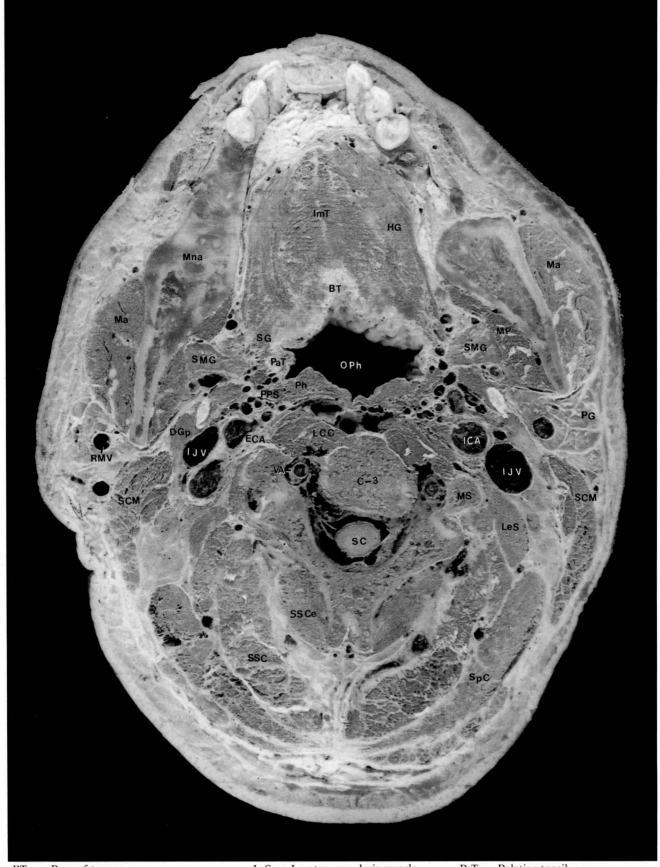

BT	Base of tongue	LeS	Levator scapularis muscle	PaT	Palatine tonsil
C3	C3 vertebral body	MP	Medial pterygoid muscle	Ph	Pharyngeal muscle
DGp	Digastric muscle, posterior belly	MS	Middle scalene muscle	RMV	Retromandibular vein
ECA	External carotid artery	Ma	Masseter muscle	SC	Spinal cord
HG	Hyoglossus muscle	Mna	Mandible, angle	SCM	Sternocleiodomastoid muscle
ICA	Internal carotid artery	Mnb	Mandible, body	SG	Styloglossus muscle
IJV	Internal jugular vein	OPh	Oropharynx	SMG	Submandibular gland
ImT	Intrinsic muscle of tongue	PG	Parotid gland	SSC	Semispinalis capitis muscle
LCC	Longus capitis & colli muscle	PPS	Parapharyngeal space	SSCe	Semispinalis cervicalis muscle

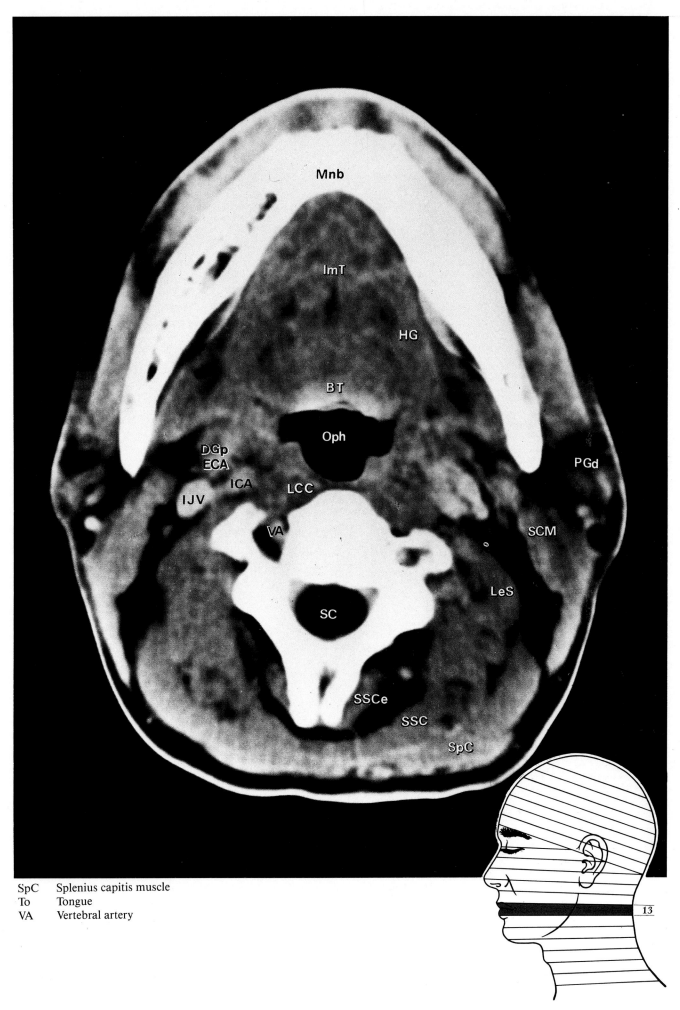

SpC   Splenius capitis muscle
To   Tongue
VA   Vertebral artery

Plate 14. TRANSVERSE HEAD AND NECK

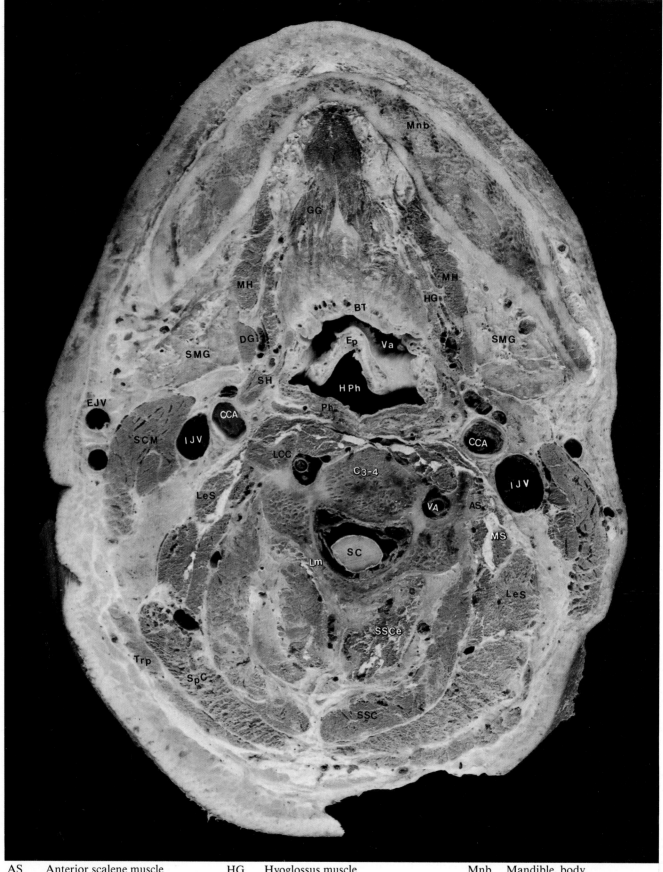

AS	Anterior scalene muscle	HG	Hyoglossus muscle	Mnb	Mandible, body
BT	Base of tongue	HPh	Hypopharynx	PPS	Parapharyngeal space
C3-4	C3-4 intervertebral disc	ICA	Internal carotid artery	Ph	Pharyngeal muscle
CCA	Common carotid artery	IJV	Internal jugular vein	SC	Spinal cord
DG	Digastric muscle	LCC	Longus capitis & colli muscle	SCM	Sternocleidomastoid muscle
ECA	External carotid artery	LeS	Levator scapularis muscle	SH	Stylohyoid muscle
EJV	External jugular vein	Lm	Lamina	SMG	Submandibular gland
Ep	Epiglottis	MH	Mylohyoid muscle	SSC	Semispinalis capitis muscle
GG	Genioglossus muscle	MS	Middle scalene muscle	SSCe	Semispinalis cervicalis muscle

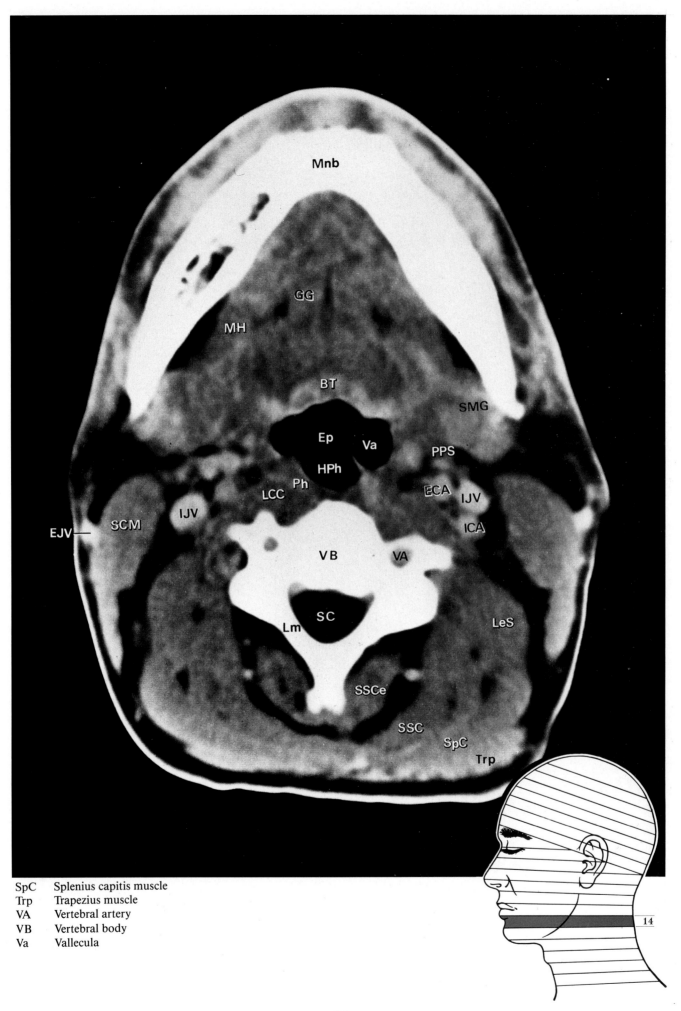

SpC   Splenius capitis muscle
Trp   Trapezius muscle
VA   Vertebral artery
VB   Vertebral body
Va   Vallecula

Plate 15. TRANSVERSE HEAD AND NECK

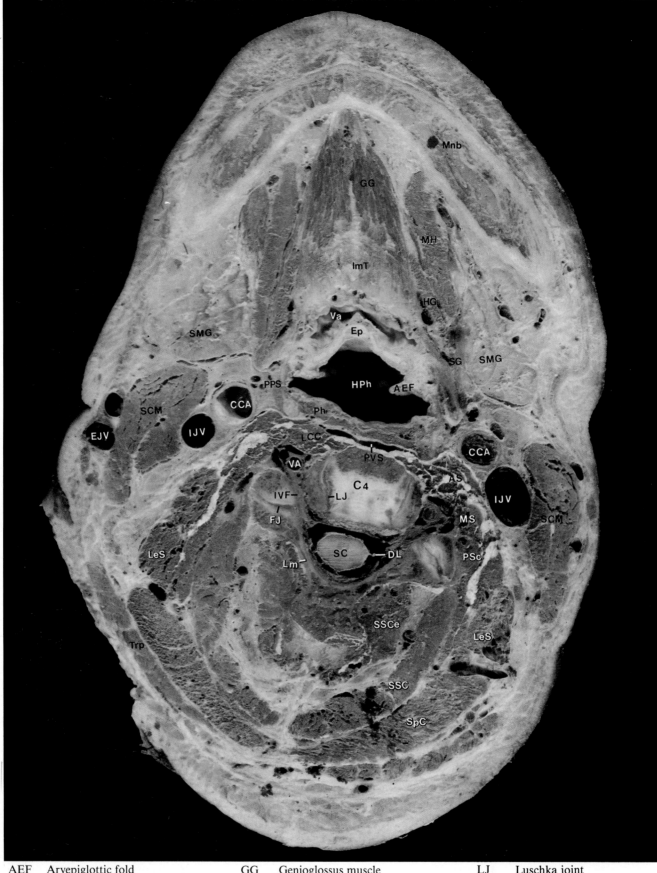

AEF	Aryepiglottic fold	GG	Genioglossus muscle	LJ	Luschka joint
AS	Anterior scalene muscle	HB	Hyoid bone	LeS	Levator scapularis muscle
C4	C4 vertebral body	HG	Hyoglossus muscle	Lm	Lamina
CCA	Common carotid artery	Hph	Hypopharynx	MH	Mylohyoid muscle
DL	Dentate ligament	ICA	Internal carotid artery	MS	Middle scalene muscle
ECA	External carotid artery	IJV	Internal jugular vein	Mnb	Mandible, body
EJV	External jugular vein	IVF	Intervertebral foramen	PPS	Parapharyngeal space
Ep	Epiglottis	ImT	Intrinsic muscle of tongue	PSc	Posterior scalene muscle
FJ	Facet joint of cervical spine	LCC	Longus capitis & colli muscle	PVS	Prevertebral space

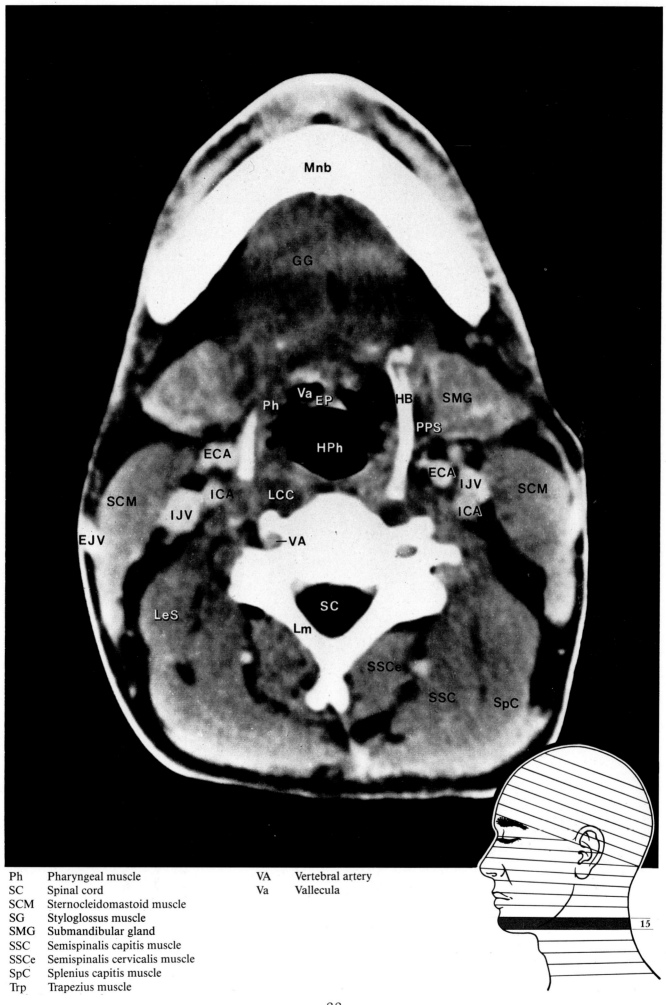

Ph	Pharyngeal muscle	VA	Vertebral artery
SC	Spinal cord	Va	Vallecula
SCM	Sternocleidomastoid muscle		
SG	Styloglossus muscle		
SMG	Submandibular gland		
SSC	Semispinalis capitis muscle		
SSCe	Semispinalis cervicalis muscle		
SpC	Splenius capitis muscle		
Trp	Trapezius muscle		

Plate 16. TRANSVERSE HEAD AND NECK

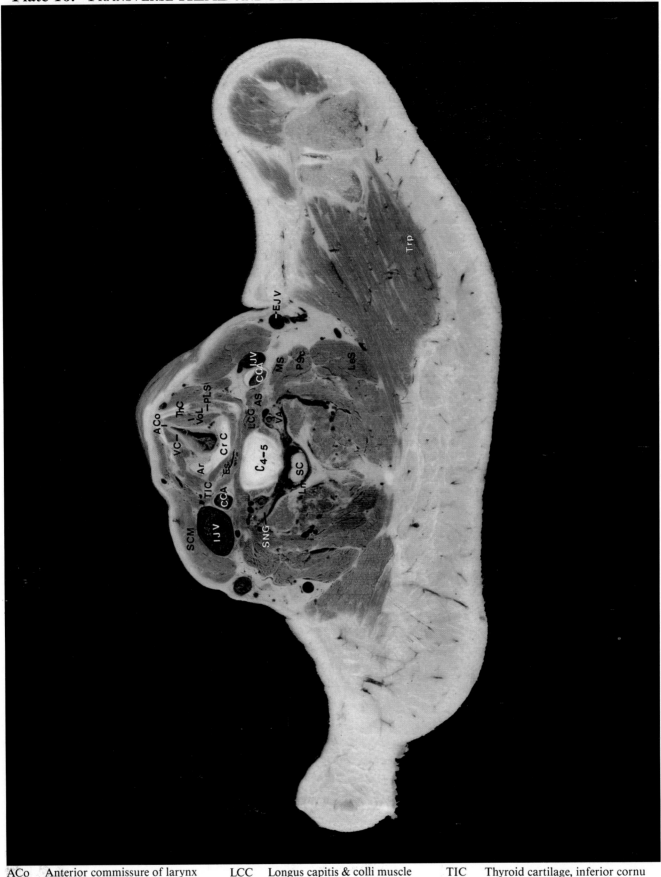

ACo	Anterior commissure of larynx	LCC	Longus capitis & colli muscle	TIC	Thyroid cartilage, inferior cornu
AS	Anterior scalene muscle	LeS	Levator scapularis muscle	ThC	Thyroid cartilage
Ar	Arytenoid	Lm	Lamina	TrF	Transverse foramen
C4-5	C4-5 intervertebral disc	MS	Middle scalene muscle	Trp	Trapezius muscle
CCA	Common carotid artery	PLS	Paralaryngeal space	VA	Vertebral artery
CrC	Cricoid cartilage	PSc	Posterior scalene muscle	VB	Vertebral body
EJV	External jugular vein	SC	Spinal cord	VC	Vocal cord
Es	Esophagus	SCM	Sternocleidomastoid muslce	VoL	Vocal ligament
IJV	Internal jugular vein	SNG	Spinal nerve ganglion		

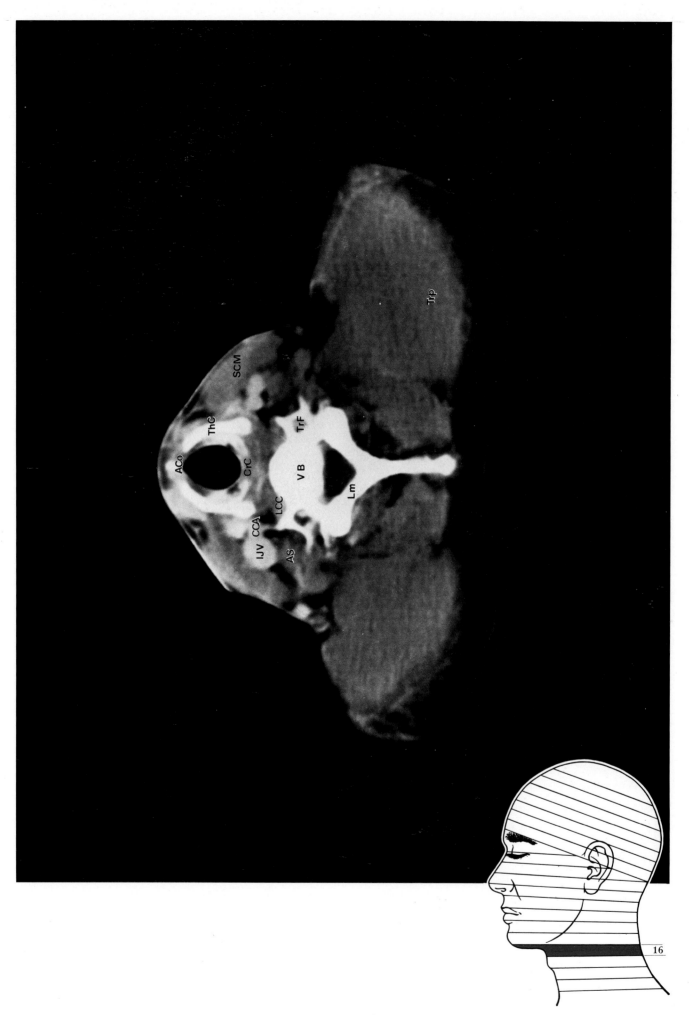

Plate 17. TRANSVERSE HEAD AND NECK

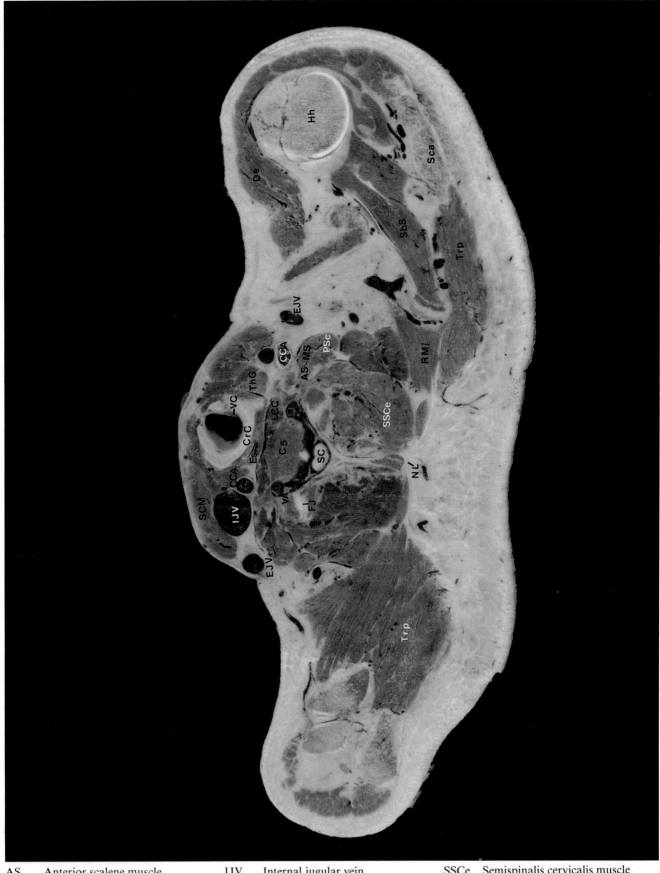

AS	Anterior scalene muscle	IJV	Internal jugular vein	SSCe	Semispinalis cervicalis muscle
C5	C5 vertebral body	LCC	Longus capitis & colli muscle	SbS	Subscapular muscle
CCA	Common carotid artery	MS	Middle scalene muscle	Sca	Scapula
CrC	Cricoid cartilage	NL	Nuchal ligament	ThG	Thyroid gland
De	Deltoid muscle	PSc	Posterior scalene muscle	Tr	Trachea
EJV	External jugular vein	RMi	Rhomboideus minor muscle	TrF	Transverse foramen
Es	Esophagus	SC	Spinal cord	Trp	Trapezius muscle
FJ	Facet joint of cervical spine	SCM	Sternocleidomastoid muscle	VA	Vertebral artery
Hh	Humerus, head	SCe	Spinalis cervicalis muscle	VC	Vocal cord

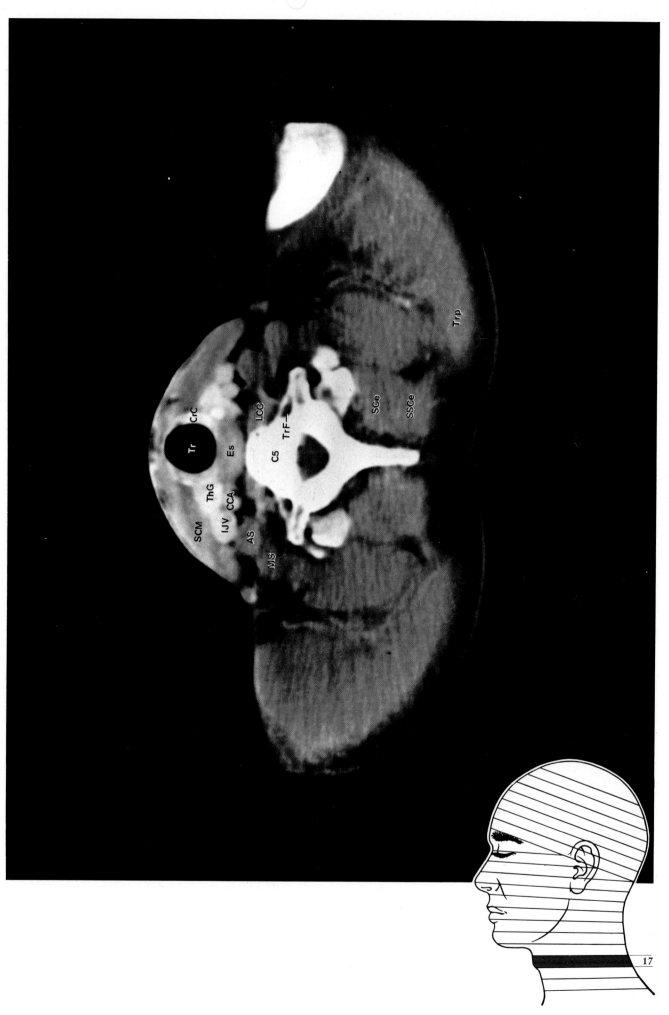

Plate 18. TRANSVERSE HEAD AND NECK

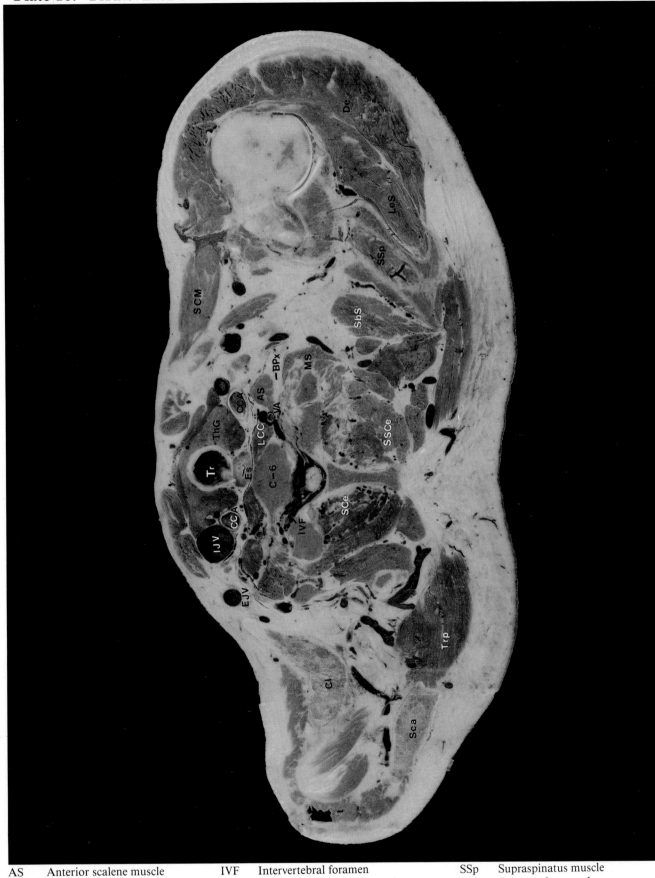

AS	Anterior scalene muscle	IVF	Intervertebral foramen	SSp	Supraspinatus muscle
BPx	Brachial plexus	LCC	Longus capitis & colli muscle	SbS	Subscapular muscle
C6	C6 vertebral body	LeS	Levator scapularis muscle	Sca	Scapula
CCA	Common carotid artery	MS	Middle scalene muscle	ThG	Thyroid gland
Cl	Clavicle	PSc	Posterior scalene muscle	Tr	Trachea
De	Deltoid muscle	SC	Spinal cord	Trp	Trapezius muscle
EJV	External jugular vein	SCM	Sternocleidomastoid muscle	VA	Vertebral artery
Es	Esophagus	SCe	Spinalis cervicalis muscle		
IJV	Internal jugular vein	SSCe	Semispinalis cervicalis muscle		

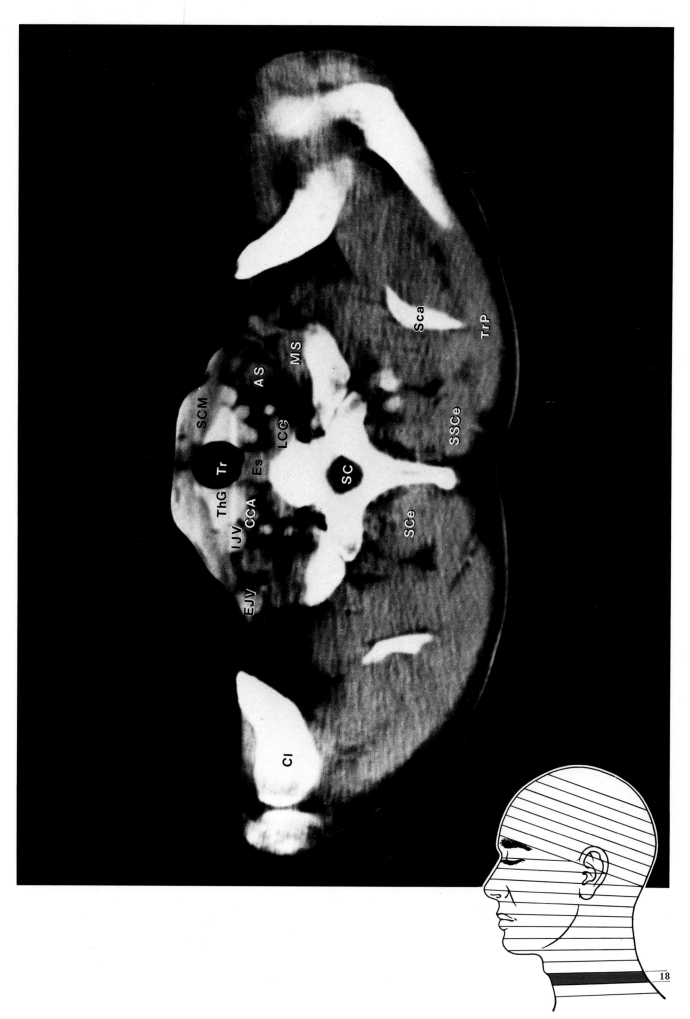

Plate 19. TRANSVERSE HEAD AND NECK

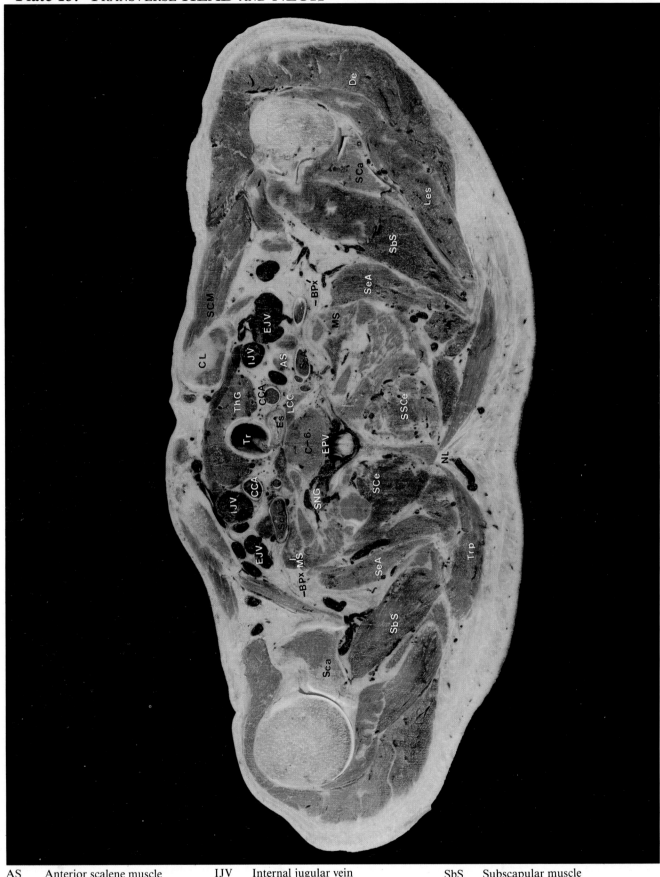

AS	Anterior scalene muscle	IJV	Internal jugular vein	SbS	Subscapular muscle
BPx	Brachial plexus	LCC	Longus capitis & colli muscle	ScA	Subclavian artery
C6	C6 vertebral body	LeS	Levator scapularis muscle	Sca	Scapula
CCA	Common carotid artery	MS	Middle scalene muscle	SeA	Serratus anterior muscle
Cl	Clavicle	NL	Nuchal ligament	ThG	Thyroid gland
De	Deltoid muscle	SCM	Sternocleidomastoid muscle	Tr	Trachea
EJV	External jugular vein	SCe	Spinalis cervicalis muscle	Trp	Trapezius muscle
EPV	Epidural venous plexus	SNG	Spinal nerve ganglion		
Es	Esophagus	SSCe	Semispinalis cervicalis muscle		

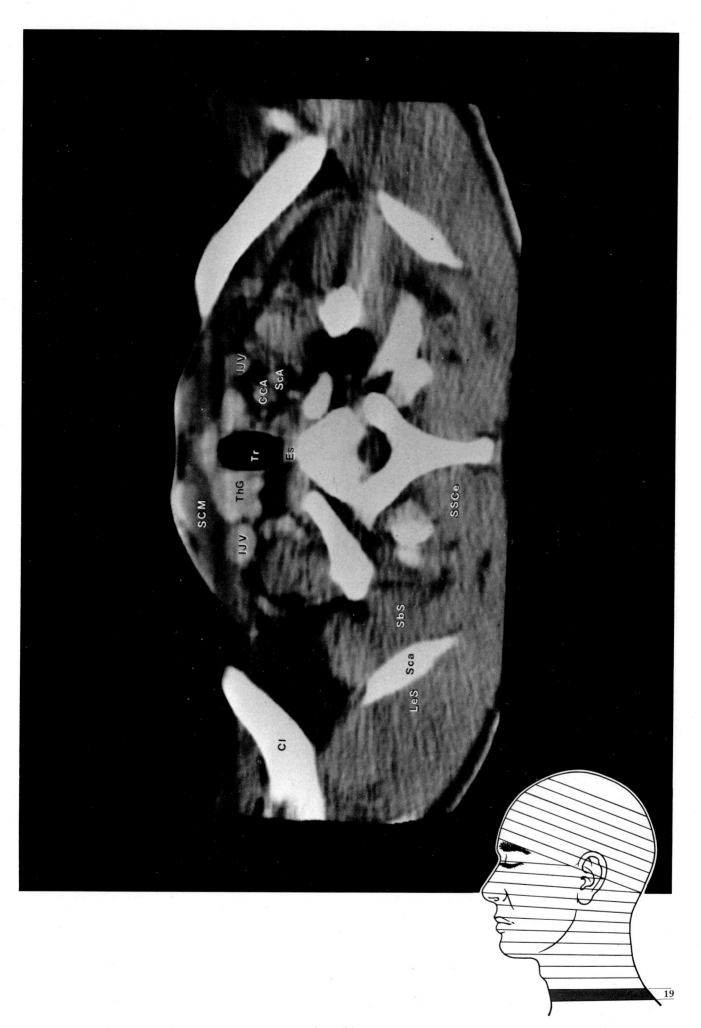

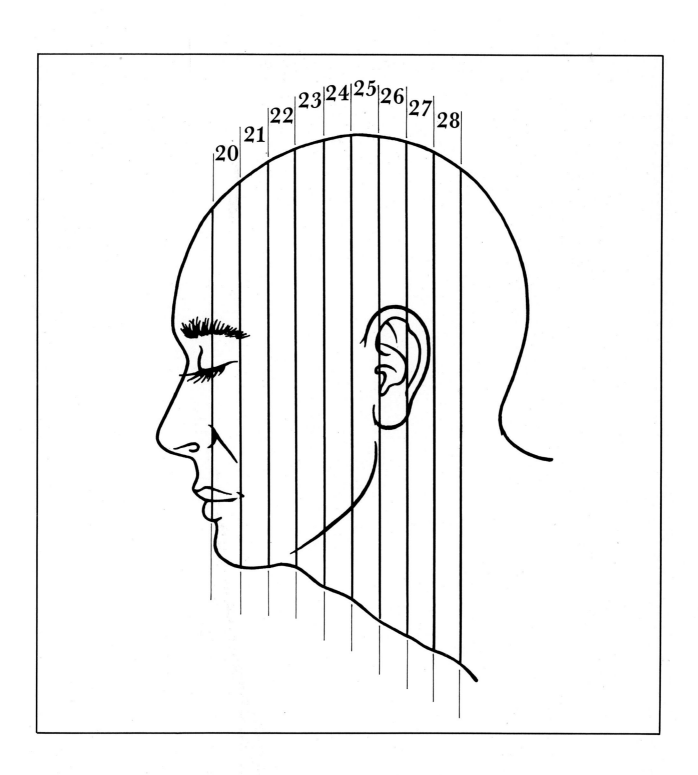

Coronal Section (Specimen & CT) Plate 20-28

Plate 20. CORONAL BRAIN, HEAD AND NECK

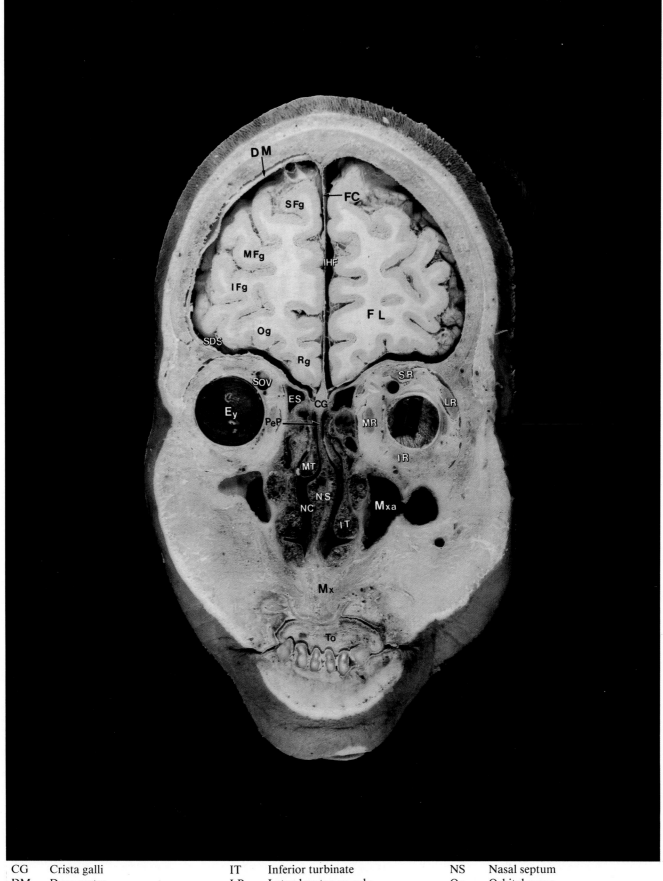

CG	Crista galli	IT	Inferior turbinate	NS	Nasal septum
DM	Dura mater	LR	Lateral rectus muscle	Og	Orbital gyrus
ES	Ethmoid sinus	LmP	Lamina papiracea	PeP	Perpendicular plate
Ey	Eyeball	MFg	Middle frontal gyrus	Rg	Rectal gyrus
FC	Falx cerebri	MR	Medial rectus muscle	SDS	Subdural space
FL	Frontal lobe	MT	Middle turbinate	SFg	Superior frontal gyrus
IFg	Inferior frontal gyrus	Mx	Maxilla	SOV	Superior opthalmic vein, dilated
IHF	Interhemispheric fissure	Mxa	Maxillary sinus, antrum	SR	Superior rectus muscle
IR	Inferior rectus muscle	NC	Nasal cavity	To	Tongue

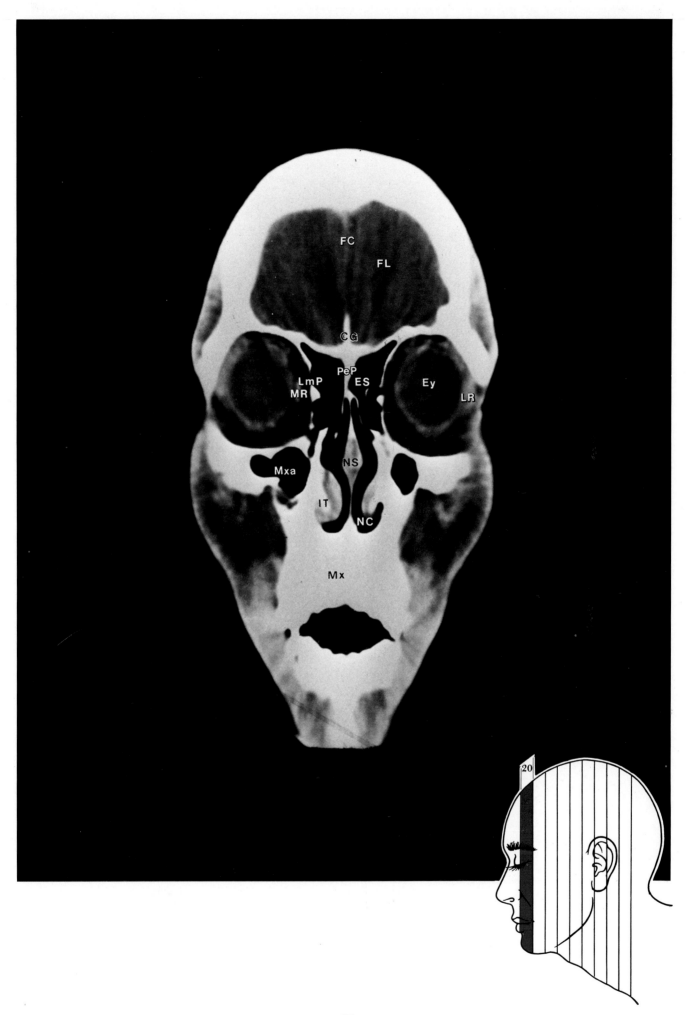

Plate 21. CORONAL BRAIN, HEAD AND NECK

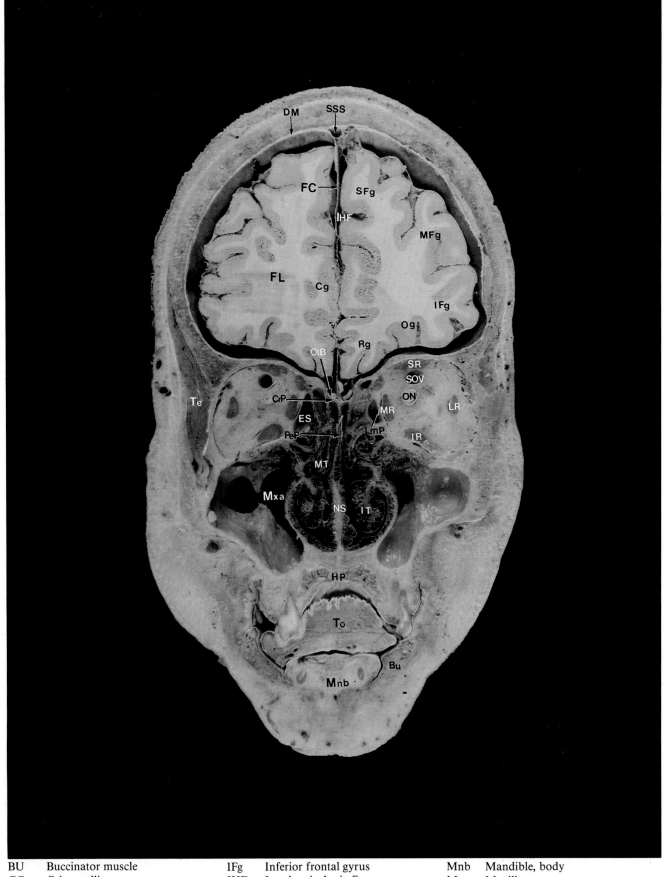

BU	Buccinator muscle	IFg	Inferior frontal gyrus	Mnb	Mandible, body
CG	Crista galli	IHF	Interhemispheric fissure	Mx	Maxilla
Cg	Cingulate gyrus	IR	Inferior rectus muscle	Mxa	Maxillary sinus, antrum
CrP	Cribriform plate	IT	Inferior turbinate	NS	Nasal septum
DM	Dura mater	LR	Lateral rectus muscle	ON	Optic nerve
ES	Ethmoid sinus	LmP	Lamina papiracea	Og	Orbital gyrus
FC	Falx cerebri	MFg	Middle frontal gyrus	OlB	Olfactory bulb
FL	Frontal lobe	MR	Medial rectus muscle	PeP	Perpendicular plate
HP	Hard palate	MT	Middle turbinate	Rg	Rectal gyrus

SFg   Superior frontal gyrus
SOV   Superior opthalmic vein, dilated
SR   Superior rectus muscle
SSS   Superior sagittal sinus
Te   Temporalis muscle
To   Tongue

Plate 22. CORONAL BRAIN, HEAD AND NECK

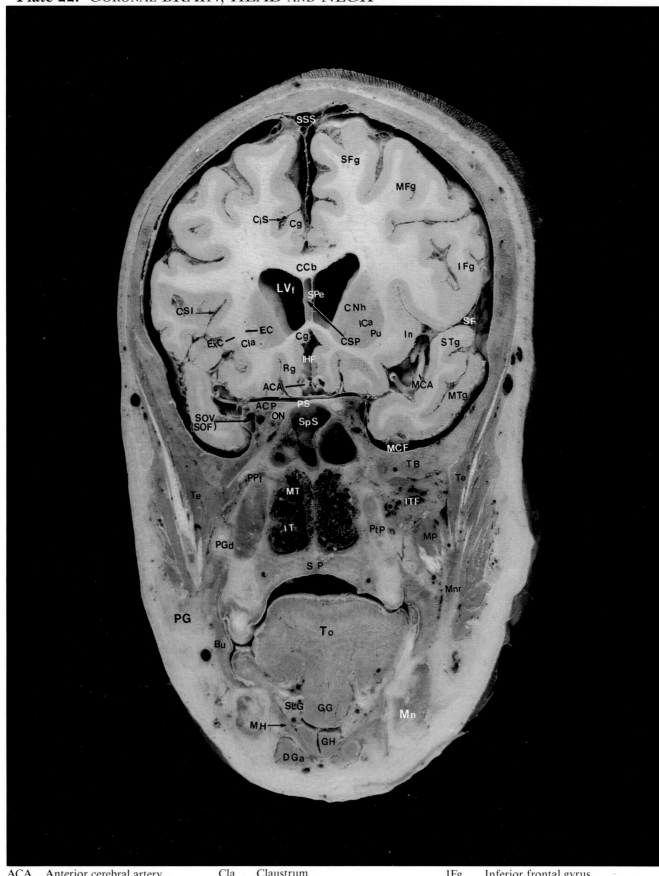

ACA	Anterior cerebral artery	Cla	Claustrum	IFg	Inferior frontal gyrus
ACP	Anterior clinoid process	DGa	Digastric muscle, anterior belly	IHF	Interhemispheric fissure
Bu	Buccinator muscle	EC	External capsule	IT	Inferior turbinate
CCb	Corpus callosum, body	ExC	Extreme capsule	ITF	Infratemporal fossa
CNh	Caudate nucleus, head	FC	Falx cerebri	In	Insula
CSI	Circular sulcus of insula	FL	Frontal lobe	LVf	Lateral ventricle, frontal horn
CSP	Cavum septum pelludicum	GG	Genioglossus muscle	MCA	Middle cerebral artery
Cg	Cingulate gyrus	GH	Geniohyoid muscle	MCF	Middle cranial fossa
CiS	Cingulate sulcus	ICa	Internal capsule, anterior limb	MFg	Middle frontal gyrus

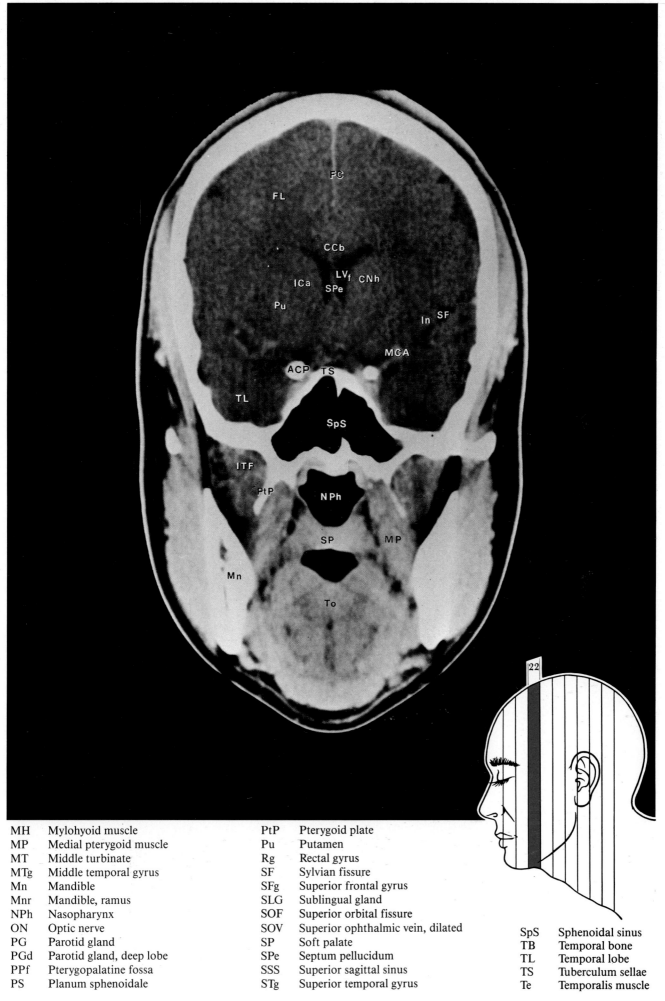

MH	Mylohyoid muscle	PtP	Pterygoid plate	
MP	Medial pterygoid muscle	Pu	Putamen	
MT	Middle turbinate	Rg	Rectal gyrus	
MTg	Middle temporal gyrus	SF	Sylvian fissure	
Mn	Mandible	SFg	Superior frontal gyrus	
Mnr	Mandible, ramus	SLG	Sublingual gland	
NPh	Nasopharynx	SOF	Superior orbital fissure	
ON	Optic nerve	SOV	Superior ophthalmic vein, dilated	
PG	Parotid gland	SP	Soft palate	SpS	Sphenoidal sinus
PGd	Parotid gland, deep lobe	SPe	Septum pellucidum	TB	Temporal bone
PPf	Pterygopalatine fossa	SSS	Superior sagittal sinus	TL	Temporal lobe
PS	Planum sphenoidale	STg	Superior temporal gyrus	TS	Tuberculum sellae
				Te	Temporalis muscle

Plate 23. CORONAL BRAIN, HEAD AND NECK

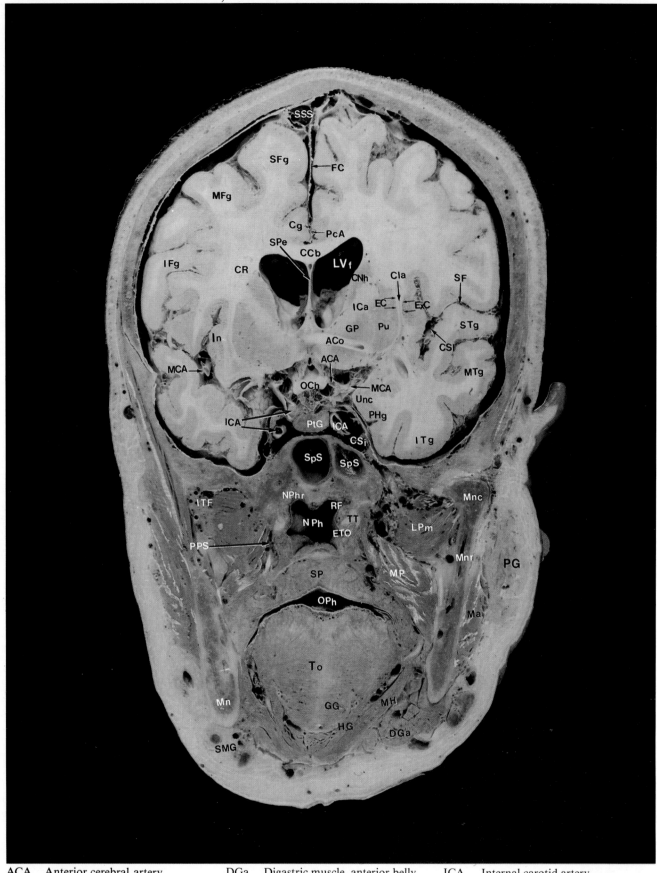

ACA	Anterior cerebral artery	DGa	Digastric muscle, anterior belly	ICA	Internal carotid artery
ACo	Anterior commissure	EC	External capsule	ICa	Internal capsule, anterior limb
CCb	Corpus callosum, body	ETO	Eustachian tube orifice	IFg	Inferior frontal gyrus
CNh	Caudate nucleus, head	ExC	Extreme capsule	IHF	Interhemispheric fissure
CR	Corona radiata	FC	Falx cerebri	ITF	Infratemporal fossa
CSI	Circular sulcus of insula	FL	Frontal lobe	ITg	Inferior temporal gyrus
CSi	Cavernous sinus	GG	Genioglossus muscle	In	Insula
Cg	Cingulate gyrus	GP	Globus pallidus	LPm	Lateral pterygoid muscle
Cla	Claustrum	HG	Hyoglossus muscle	LVf	Lateral ventricle, frontal horn

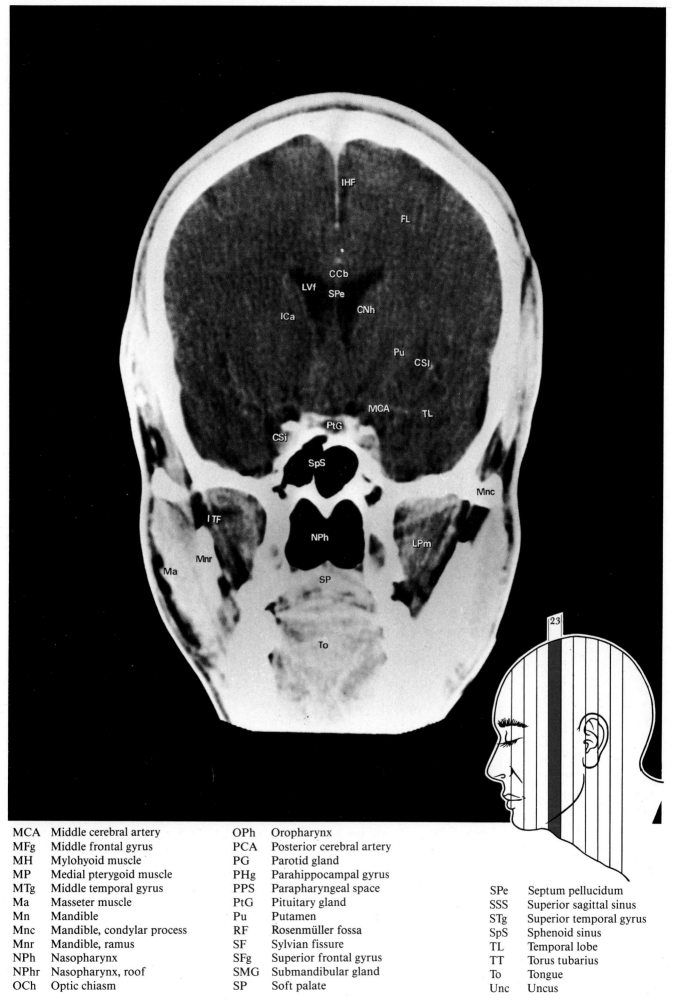

MCA	Middle cerebral artery	OPh	Oropharynx	
MFg	Middle frontal gyrus	PCA	Posterior cerebral artery	
MH	Mylohyoid muscle	PG	Parotid gland	
MP	Medial pterygoid muscle	PHg	Parahippocampal gyrus	
MTg	Middle temporal gyrus	PPS	Parapharyngeal space	
Ma	Masseter muscle	PtG	Pituitary gland	SPe	Septum pellucidum
Mn	Mandible	Pu	Putamen	SSS	Superior sagittal sinus
Mnc	Mandible, condylar process	RF	Rosenmüller fossa	STg	Superior temporal gyrus
Mnr	Mandible, ramus	SF	Sylvian fissure	SpS	Sphenoid sinus
NPh	Nasopharynx	SFg	Superior frontal gyrus	TL	Temporal lobe
NPhr	Nasopharynx, roof	SMG	Submandibular gland	TT	Torus tubarius
OCh	Optic chiasm	SP	Soft palate	To	Tongue
				Unc	Uncus

Plate 24. CORONAL BRAIN, HEAD AND NECK

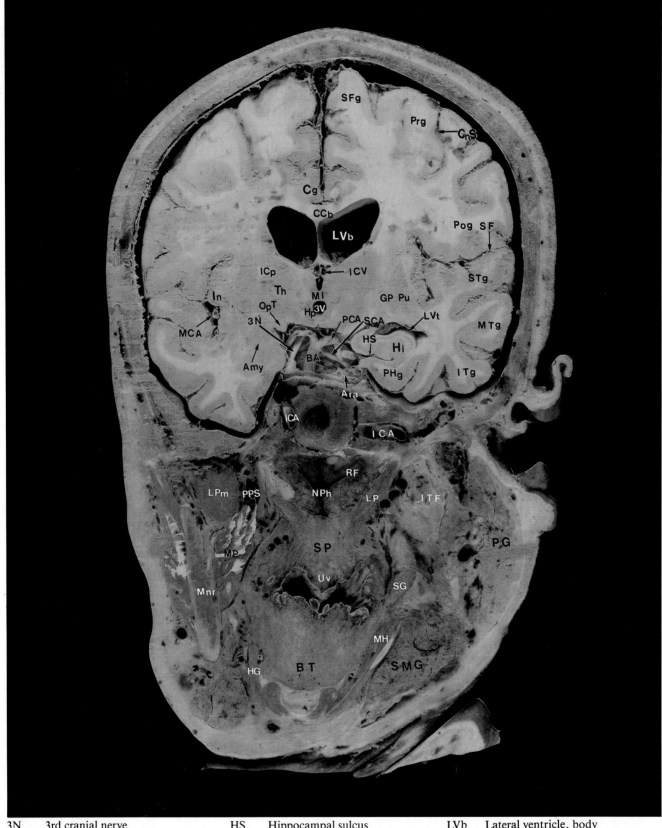

3N	3rd cranial nerve	HS	Hippocampal sulcus	LVb	Lateral ventricle, body
3V	3rd ventricle	Hi	Hippocampus	LVf	Lateral ventricle, frontal horn
Amy	Amygdala	Hp	Hypothalamus	LVt	Lateral ventricle, temporal horn
Ara	Arachnoid membrane	ICA	Internal carotid artery	MCA	Middle cerebral artery
BA	Basilar artery	ICV	Internal cerebral vein	MH	Mylohyoid muscle
BT	Base of tongue	ICa	Internal capsule, anterior limb	MI	Massa intermedia
CCb	Corpus callosum, body	ICp	Internal capsule, posterior limb	MP	Medial pterygoid muscle
CNh	Caudate nucleus, head	ITF	Infratemporal fossa	MTg	Middle temporal gyrus
Cg	Cingulate gyrus	ITg	Inferior temporal gyrus	Mnr	Mandible, ramus
CnS	Central sulcus	In	Insula	NPh	Nasopharynx
GP	Globus pallidus	LP	Levator palati muscle	Nphr	Nasopharynx, roof
HG	Hyoglossus muscle	LPm	Lateral pterygoid muscle	OpT	Optic tract

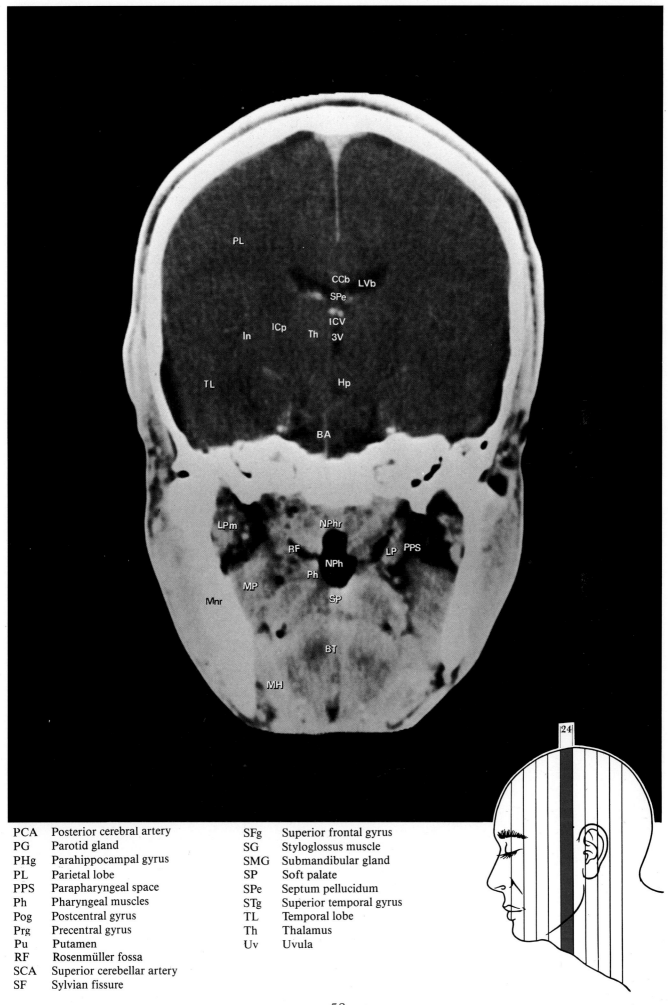

PCA	Posterior cerebral artery	SFg	Superior frontal gyrus
PG	Parotid gland	SG	Styloglossus muscle
PHg	Parahippocampal gyrus	SMG	Submandibular gland
PL	Parietal lobe	SP	Soft palate
PPS	Parapharyngeal space	SPe	Septum pellucidum
Ph	Pharyngeal muscles	STg	Superior temporal gyrus
Pog	Postcentral gyrus	TL	Temporal lobe
Prg	Precentral gyrus	Th	Thalamus
Pu	Putamen	Uv	Uvula
RF	Rosenmüller fossa		
SCA	Superior cerebellar artery		
SF	Sylvian fissure		

Plate 25. CORONAL BRAIN, HEAD AND NECK

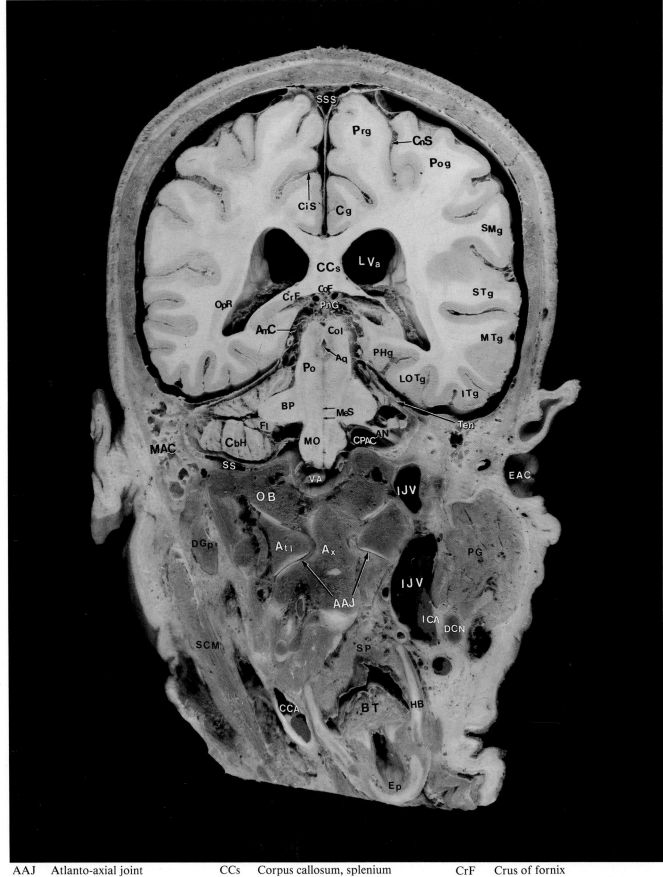

AAJ	Atlanto-axial joint	CCs	Corpus callosum, splenium	CrF	Crus of fornix
AN	Acoustic nerve	CPAC	Cerebellopontine angle cistern	DCN	Deep cervical lymph node
AmC	Ambient cistern	CbH	Cerebellar hemisphere	DGp	Digastric muscle, posterior belly
Aq	Cerebral aqueduct	Cg	Cingulate gyrus	EAC	External auditory canal
Atl	Atlas	ChP	Choroid plexus of lateral ventricle	Ep	Epiglottis
Ax	Axis	CiS	Cingulate sulcus	Fl	Flocculus
BP	Brachium pontis	CnS	Central sulcus	HB	Hyoid bone
BT	Base of tongue	CoF	Commissure of fornix	ICA	Internal carotid artery
CCA	Common carotid artery	Col	Colliculi	ICV	Internal cerebral vein

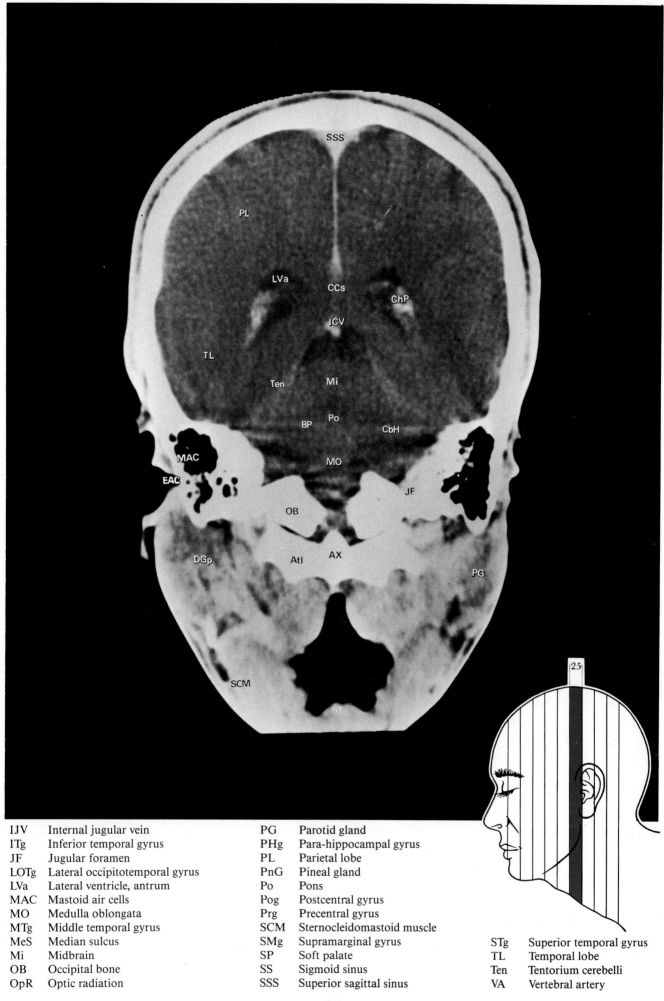

IJV	Internal jugular vein	PG	Parotid gland
ITg	Inferior temporal gyrus	PHg	Para-hippocampal gyrus
JF	Jugular foramen	PL	Parietal lobe
LOTg	Lateral occipitotemporal gyrus	PnG	Pineal gland
LVa	Lateral ventricle, antrum	Po	Pons
MAC	Mastoid air cells	Pog	Postcentral gyrus
MO	Medulla oblongata	Prg	Precentral gyrus
MTg	Middle temporal gyrus	SCM	Sternocleidomastoid muscle
MeS	Median sulcus	SMg	Supramarginal gyrus
Mi	Midbrain	SP	Soft palate
OB	Occipital bone	SS	Sigmoid sinus
OpR	Optic radiation	SSS	Superior sagittal sinus

STg	Superior temporal gyrus
TL	Temporal lobe
Ten	Tentorium cerebelli
VA	Vertebral artery

Plate 26. CORONAL BRAIN, HEAD AND NECK

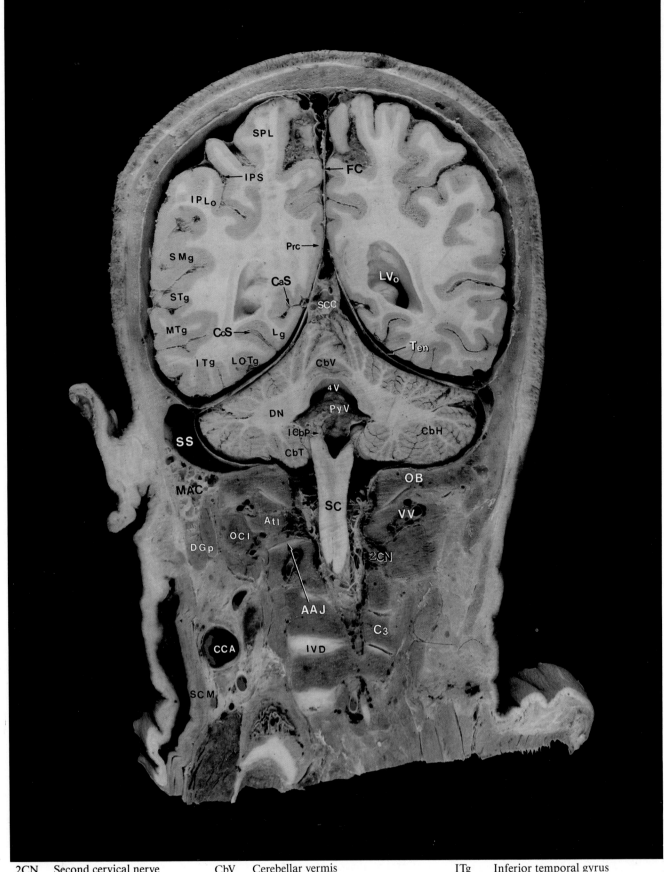

2CN	Second cervical nerve	CbV	Cerebellar vermis	ITg	Inferior temporal gyrus
4V	4th ventricle	ChP	Choroid plexus of lateral ventricle	IVD	Intervertebral disc
AAJ	Atlantoaxial joint	CoS	Collateral sulcus	LOTg	Lateral occipitotemporal gyrus
Atl	Atlas	DGp	Digastric muscle, posterior belly	LVo	Lateral ventricle, occipital horn
C3	C-3 vertebra	DN	Dentate nucleus	Lg	Lingual gyrus
CCA	Common carotid artery	FC	Falx cerebri	MAC	Mastoid air cells
CaS	Calcarine sulcus	ICbP	Inferior cerebellar peduncle	MO	Medulla oblongata
CbH	Cerebellar hemispherre	IPLo	Inferior parietal lobule	MTg	Middle temporal gyrus
CbT	Cerebellar tonsil	IPS	Intraparietal sulcus	OB	Occipital bone

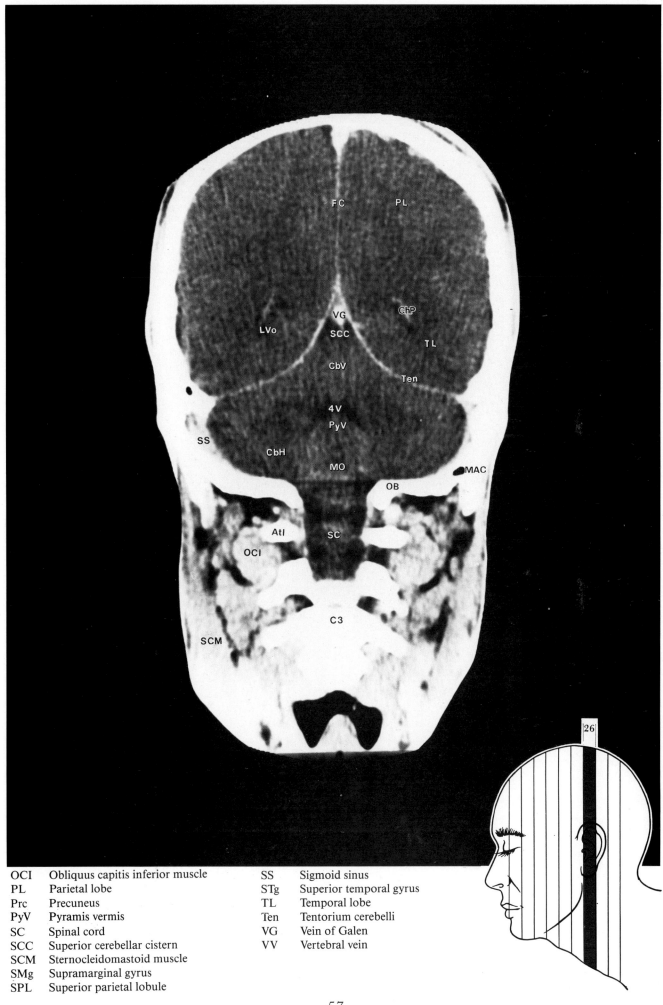

OCI	Obliquus capitis inferior muscle	SS	Sigmoid sinus
PL	Parietal lobe	STg	Superior temporal gyrus
Prc	Precuneus	TL	Temporal lobe
PyV	Pyramis vermis	Ten	Tentorium cerebelli
SC	Spinal cord	VG	Vein of Galen
SCC	Superior cerebellar cistern	VV	Vertebral vein
SCM	Sternocleidomastoid muscle		
SMg	Supramarginal gyrus		
SPL	Superior parietal lobule		

Plate 27. Coronal BRAIN, HEAD and NECK

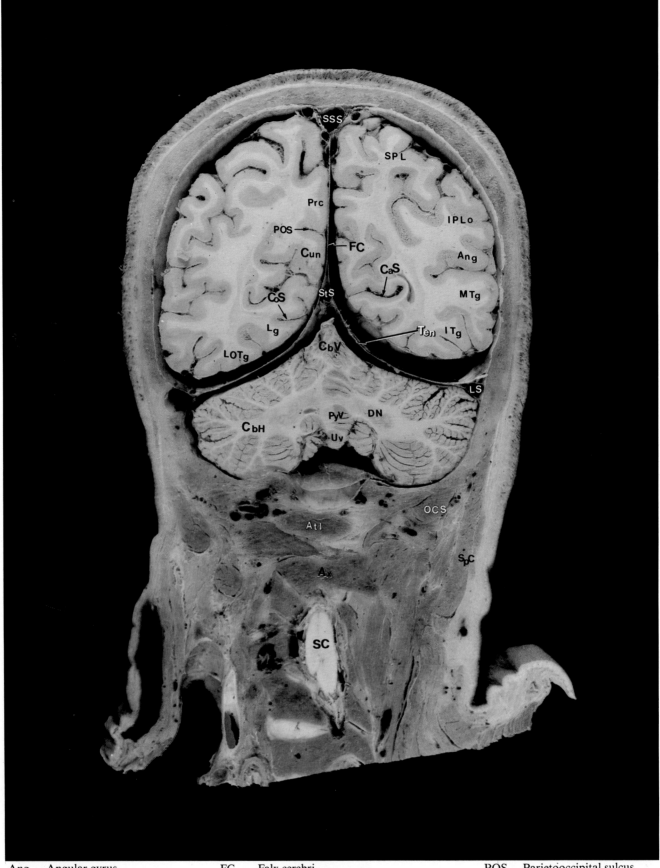

Ang	Angular gyrus	FC	Falx cerebri	POS	Parietóoccipital sulcus
Atl	Atlas	IPLo	Inferior parietal lobule	Prc	Precuneus
Ax	Axis	ITg	Inferior temporal gyrus	PyV	Pyramis vermis
CaS	Calcarine sulcus	LOTg	Lateral occipitotemporal gyrus	SC	Spinal cord
CbH	Cerebellar hemisphere	LS	Lateral sinus	SpC	Splenius capitis muscle
CbV	Cerebellar vermis	Lg	Lingual gyrus	SPL	Superior parietal lobule
CoS	Collateral sulcus	MTg	Middle temporal gyrus	SSS	Superior sagittal sinus
Cun	Cuneus	OCS	Obliquus capitis superior muscle	StS	Straight sinus
DN	Dentate nucleus	PL	Parietal lobe	TL	Temporal lobe

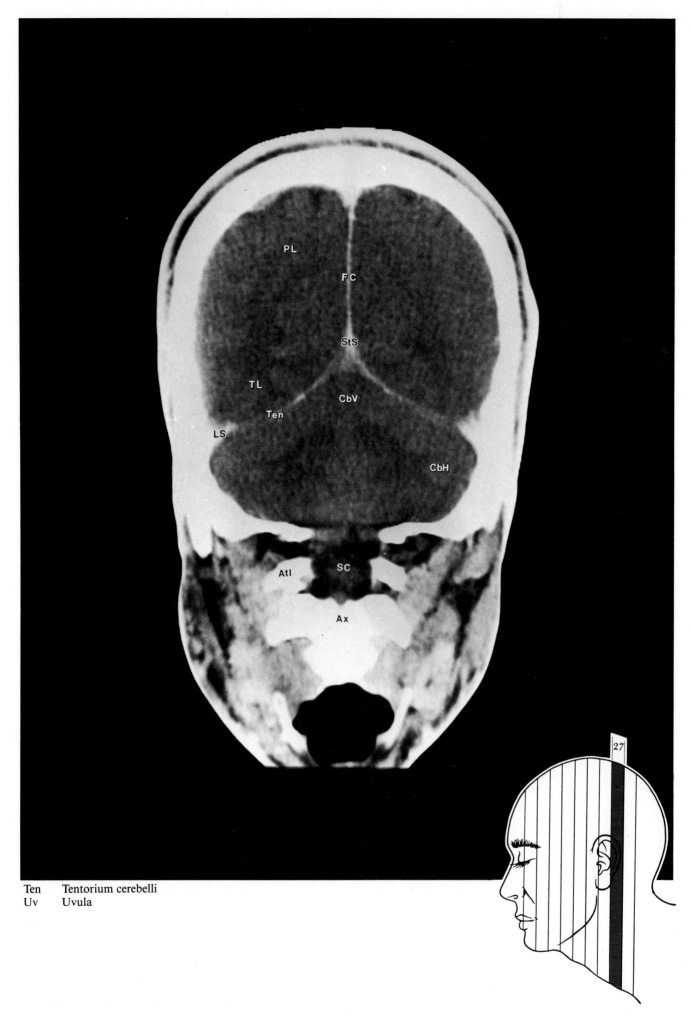

Ten   Tentorium cerebelli
Uv   Uvula

Plate 28. Coronal BRAIN, HEAD and NECK

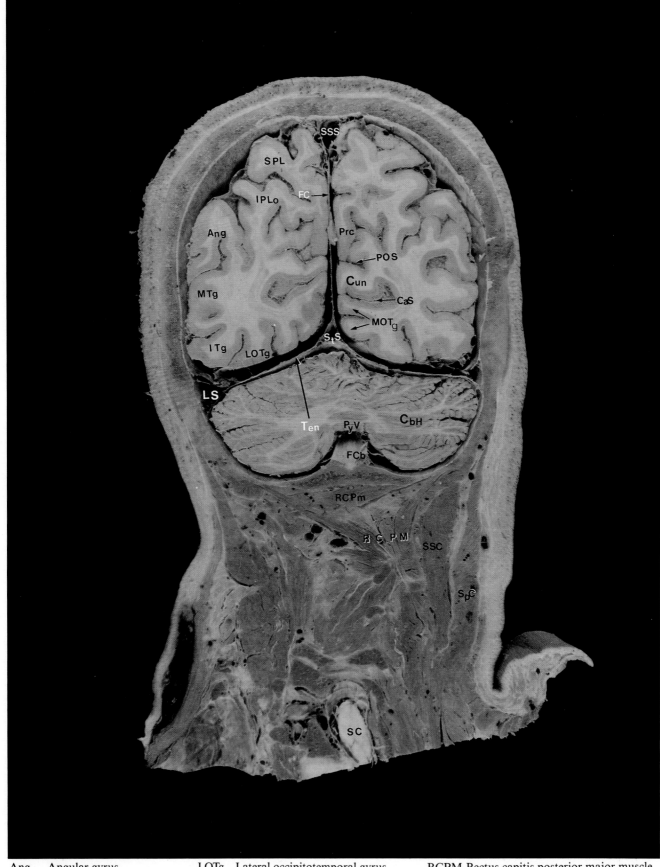

Ang	Angular gyrus	LOTg	Lateral occipitotemporal gyrus	RCPM	Rectus capitis posterior major muscle
CaS	Calcarine sulcus	LS	Lateral sinus	RCPm	Rectus capitis posterior minor muscle
CbH	Cerebellar hemisphere	MOTg	Medial occipitotemporal gyrus	SC	Spinal cord
CbV	Cerebellar vermis	MTg	Middle temporal gyrus	SPL	Superior parietal lobule
Cun	Cuneus	OB	Occipital bone	SSC	Semispinalis capitis muscle
FC	Falx cerebri	OL	Occipital lobe	SSS	Superior sagittal sinus
FCb	Falx cerebelli	POS	Parietooccipital sulcus	SpC	Splenius capitis muscle
IPLo	Inferior parietal lobule	Prc	Precuneus	StS	Straight sinus
ITg	Inferior temporal gyrus	PyV	Pyramis vermis	Ten	Tentorium cerebelli

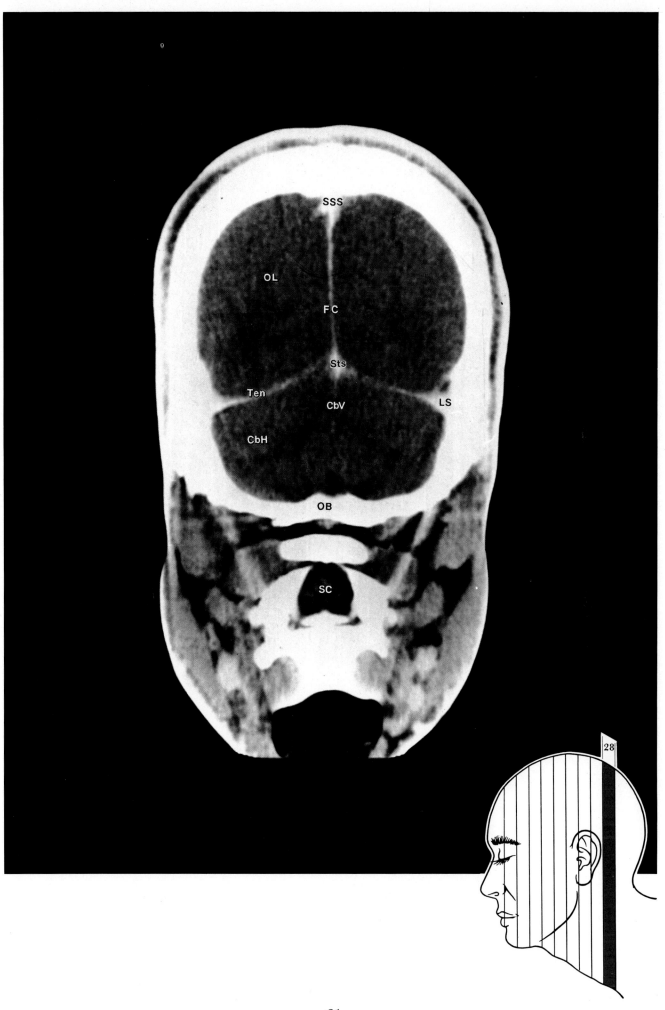

PART 2
CHEST

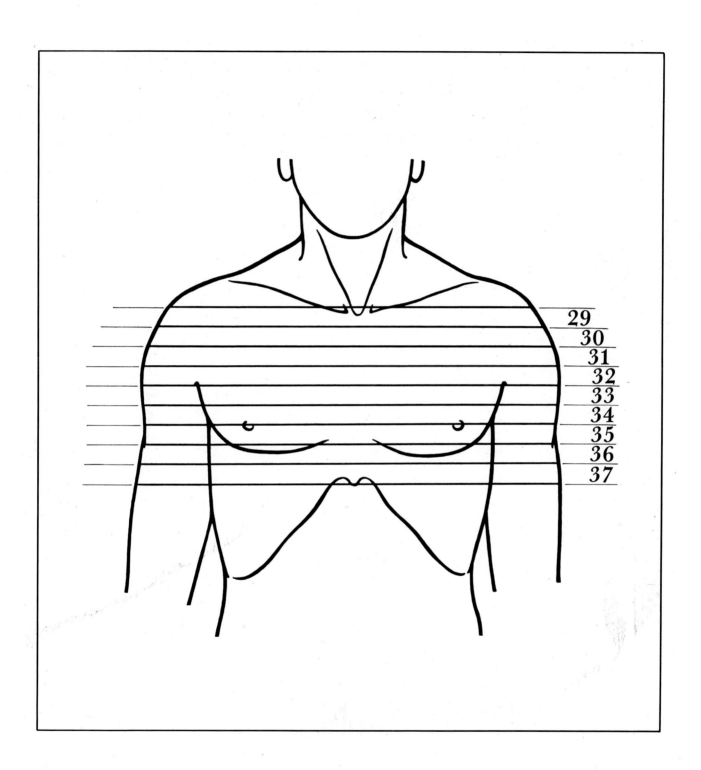

Transverse Section (Specimen & CT) Plate 29-37

Plate 29. TRANSVERSE CHEST

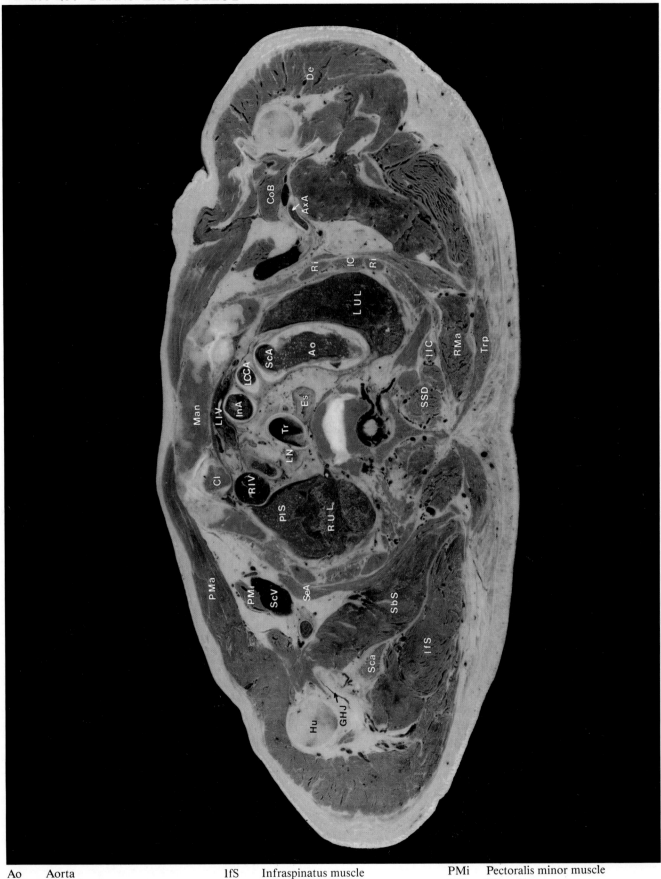

Ao	Aorta	IfS	Infraspinatus muscle
AxA	Axillary artery	IIC	Iliocostalis muscle
Cl	Clavicle	InA	Innominate artery
CoB	Coracobrachialis muscle	LCCA	Left common carotid artery
De	Deltoid muscle	LIV	Left innominate vein
Es	Esophagus	LN	Lymph node
GHJ	Glenohumeral joint	LUL	Left upper lobe of lung
Hu	Humerus	Man	Manubrium
Ic	Intercostal muscle	PMa	Pectoralis major muscle

PMi	Pectoralis minor muscle
PlS	Pleural space
RIV	Right innominate vein
RMa	Rhomboideus major muscle
RUL	Right upper lobe of lung
Ri	Rib
SSD	Semispinalis dorsi muscle
SbS	Subscapularis muscle
ScA	Subclavian artery

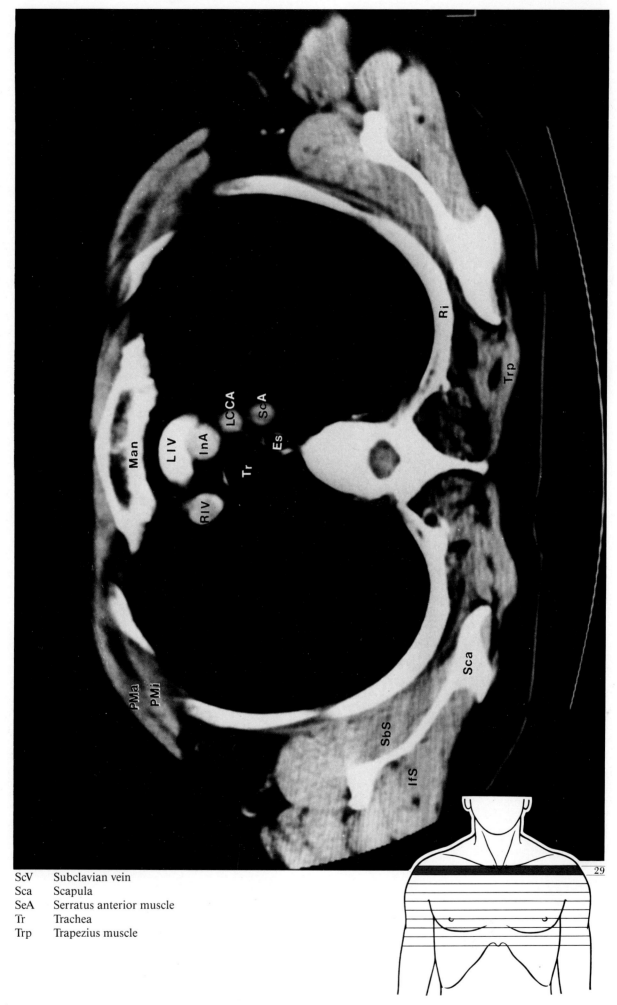

ScV Subclavian vein
Sca Scapula
SeA Serratus anterior muscle
Tr   Trachea
Trp Trapezius muscle

Plate 30. TRANSVERSE CHEST

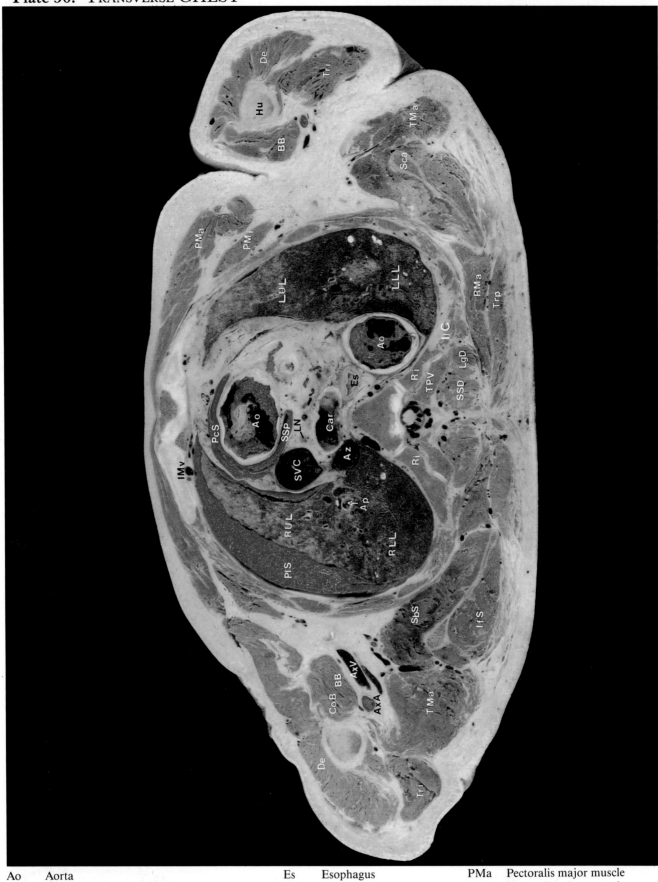

Ao	Aorta	Es	Esophagus	PMa	Pectoralis major muscle
Ap	Apical segment of right upper lobe bronchus	Hu	Humerus	PMi	Pectoralis minor muscle
AxA	Axillary artery	IMv	Internal mammary vessel	PcS	Pericardial space
AxV	Axillary vein	IfS	Infraspinatus muscle	PlS	Pleural space
Az	Azygos vein	IlC	Iliocostalis muscle	RLL	Right lower lobe of lung
BB	Biceps brachii muscle	LLL	Left lower lobe of lung	RMa	Rhomboideus major muscle
Car	Carina	LN	Lymph node	RUL	Right upper lobe of lung
CoB	Coracobrachialis muscle	LUL	Left upper lobe of lung	Ri	Rib
De	Deltoid muscle	LgD	Longissimus dorsi muscle	SSD	Semispinalis dorsi muscle

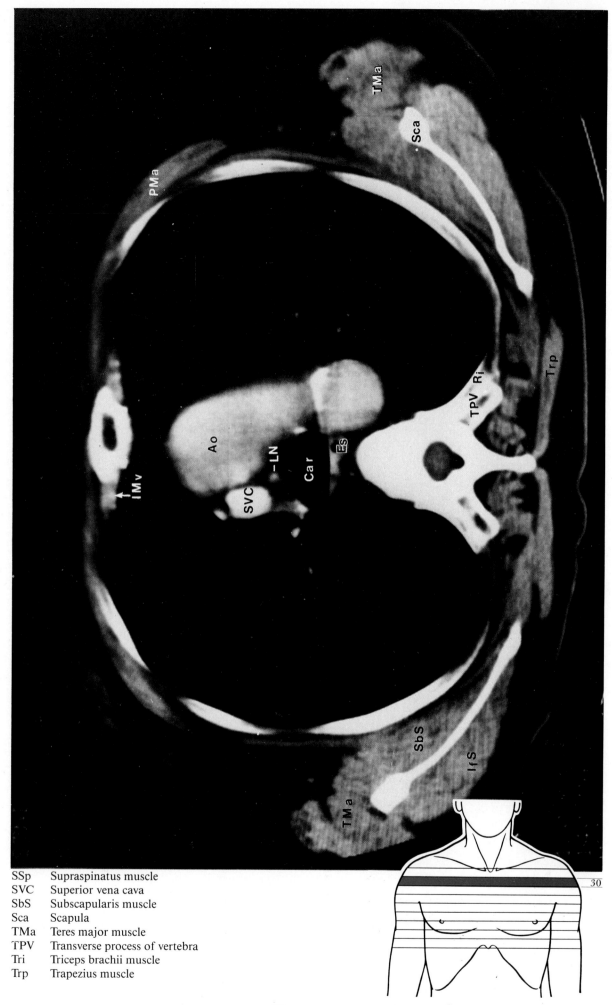

SSp   Supraspinatus muscle
SVC   Superior vena cava
SbS   Subscapularis muscle
Sca   Scapula
TMa   Teres major muscle
TPV   Transverse process of vertebra
Tri   Triceps brachii muscle
Trp   Trapezius muscle

Plate 31. TRANSVERSE CHEST

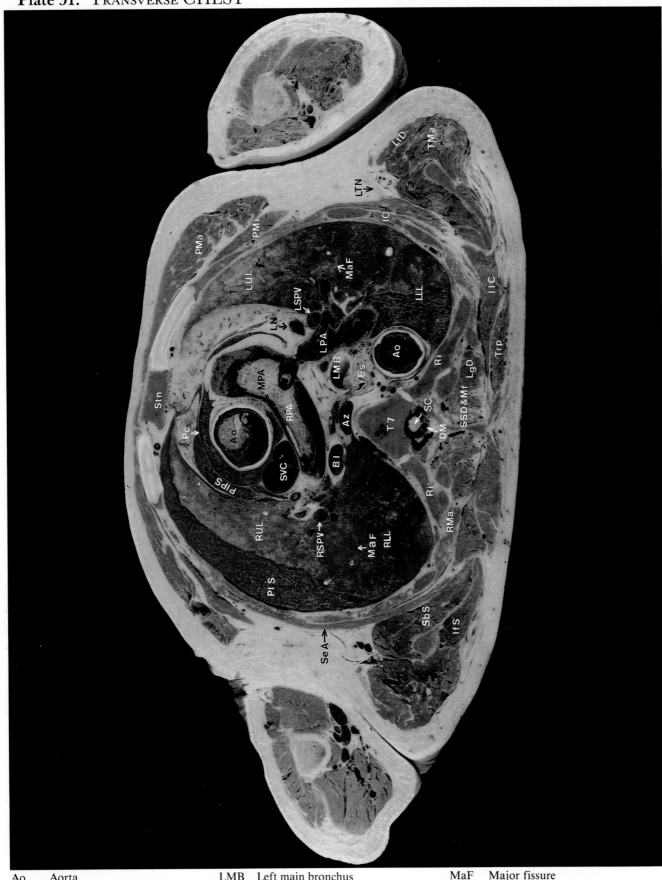

Ao	Aorta	LMB	Left main bronchus	MaF	Major fissure
Az	Azygos vein	LN	Lymph node	Mf	Multifidus muscle
BI	Bronchus intermedius	LPA	Left pulmonary artery	PMa	Pectoralis major muscle
DM	Dura mater	LSPV	Left superior pulmonary vein	PMi	Pectoralis minor muscle
Es	Esophagus	LTN	Long thoracic nerve	Pc	Pericardium
Ic	Intercostal muscle	LUL	Left upper lobe of lung	PlPS	Pleuropericardial space
IfS	Infraspinatus muscle	LgD	Longissimus dorsi muscle	PlS	Pleural space
IlC	Iliocostalis muscle	LtD	Latissimus dorsi muscle	RLL	Right lower lobe of lung
LLL	Left lower lobe of lung	MPA	Main pulmonary artery	RMa	Rhomboideus major muscle

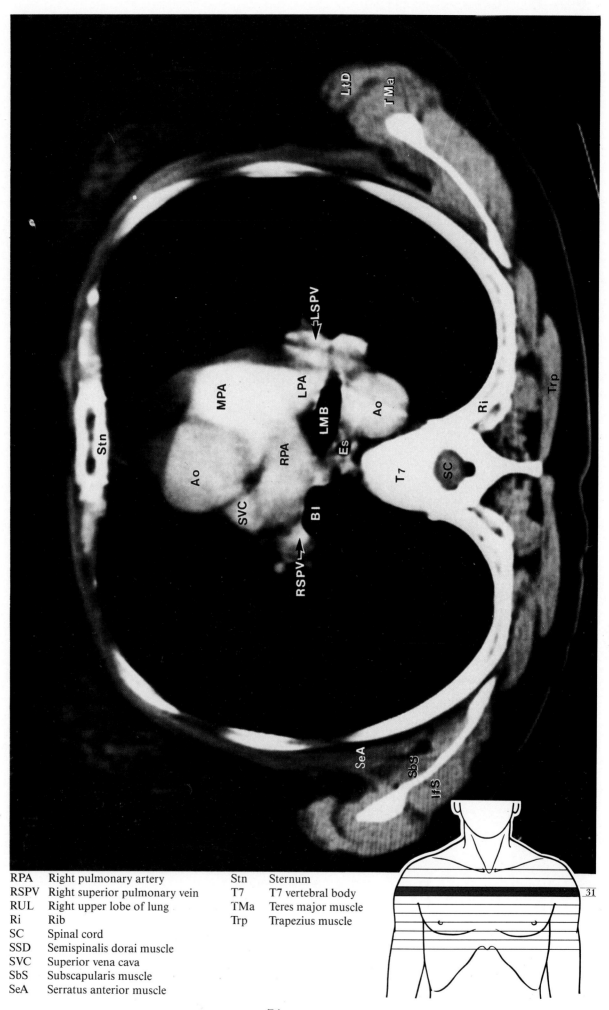

RPA	Right pulmonary artery	Stn	Sternum
RSPV	Right superior pulmonary vein	T7	T7 vertebral body
RUL	Right upper lobe of lung	TMa	Teres major muscle
Ri	Rib	Trp	Trapezius muscle
SC	Spinal cord		
SSD	Semispinalis dorai muscle		
SVC	Superior vena cava		
SbS	Subscapularis muscle		
SeA	Serratus anterior muscle		

31

Plate 32. TRANSVERSE CHEST

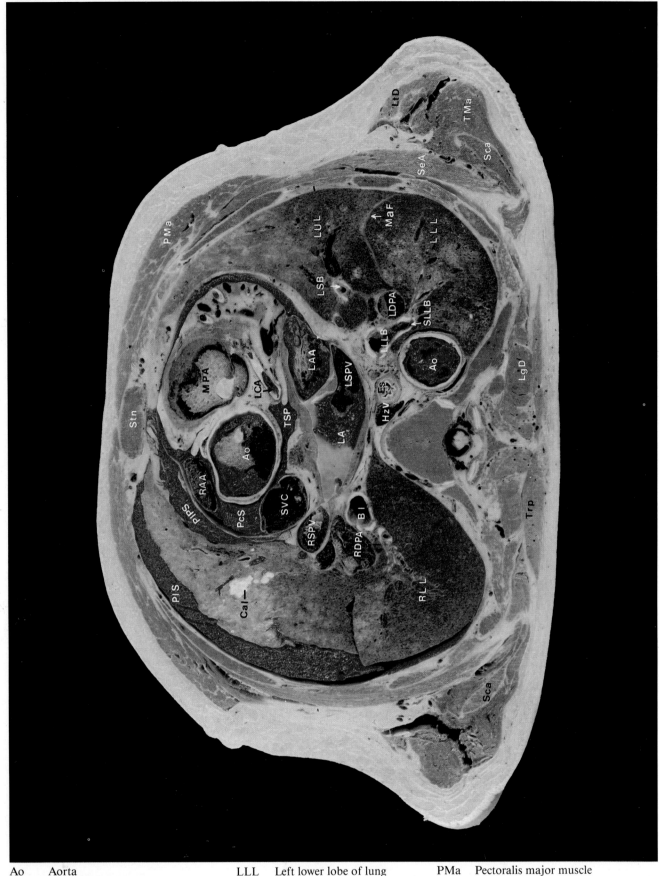

Ao	Aorta	LLL	Left lower lobe of lung	PMa	Pectoralis major muscle
BI	Bronchus intermedius	LLLB	Left lower lobe bronchus	PcS	Pericardial space
Cal	Calcification of lung	LSB	Lingular segmental bronchus	PlPS	Pleuropericardial space
Es	Esophagus	LSPV	Left superior pulmonary vein	PlS	Pleural space
HzV	Hemiazygos vein	LUL	Left upper lobe of lung	RAA	Right atrial appendage of heart
LA	Left atrium of heart	LtD	Latissimus dorsi muscle	RDPA	Right descending pulmonary artery
LAA	Left atrial appendage of heart	LgD	Longissimus dorsi muscle	RLL	Right lower lobe of lung
LCA	Left coronary artery	MPA	Main pulmonary artery	RSPV	Right superior pulmonary vein
LDPA	Left descending pulmonary artery	MaF	Major fissure	SVC	Superior vena cava

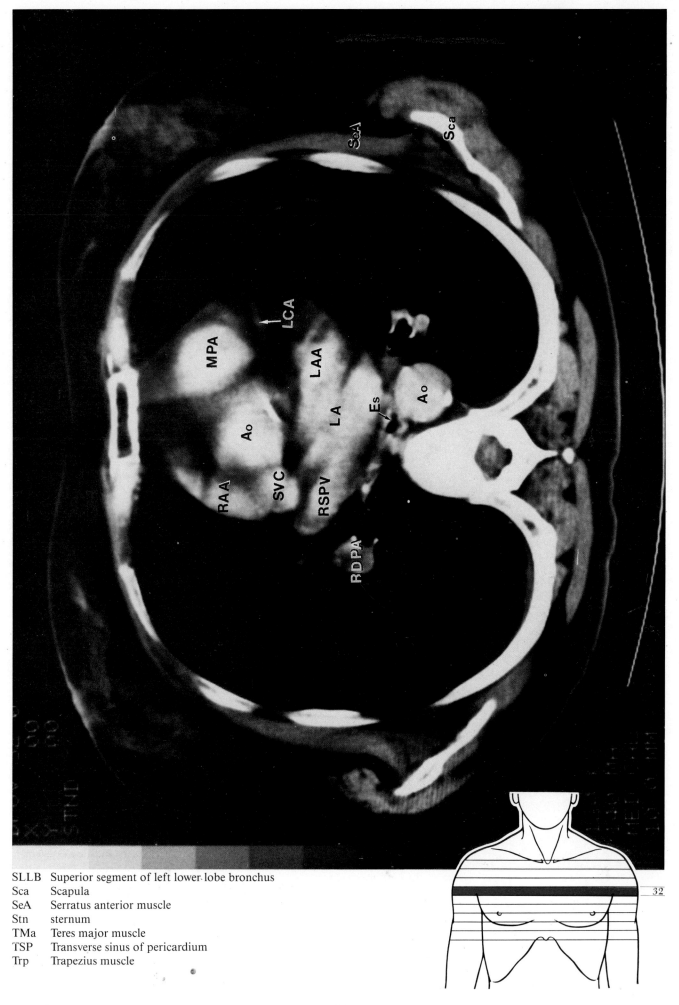

SLLB Superior segment of left lower lobe bronchus
Sca   Scapula
SeA   Serratus anterior muscle
Stn   sternum
TMa Teres major muscle
TSP   Transverse sinus of pericardium
Trp   Trapezius muscle

Plate 33. TRANSVERSE CHEST

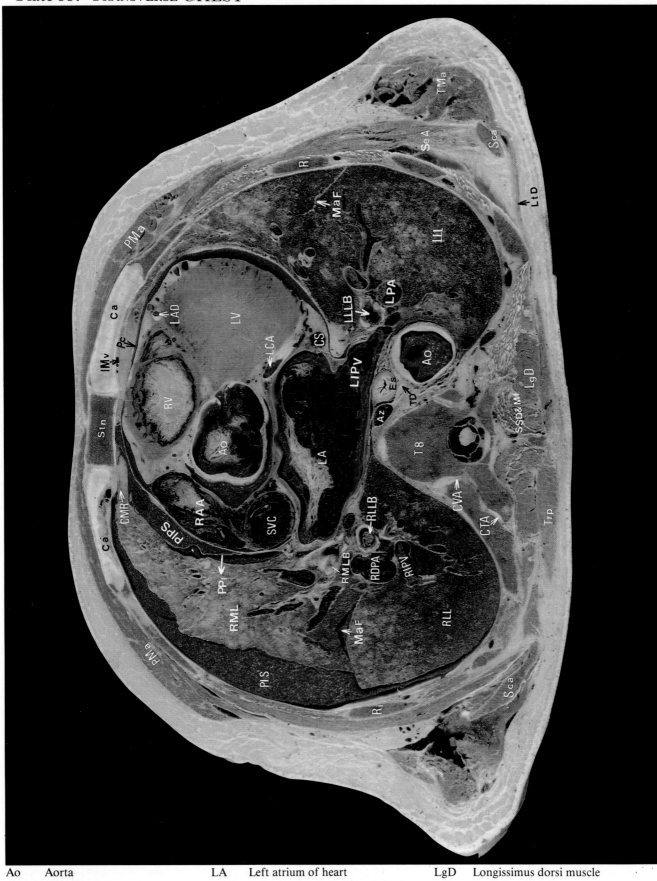

Ao	Aorta	LA	Left atrium of heart	LgD	Longissimus dorsi muscle
Az	Azygos vein	LAD	Left anterior descending branch	LtD	Latissimus dorsi muscle
CMR	Costomediastinal recess		of left coronary artery	MaF	Major fissure
CS	Coronary sinus	LCA	Left coronary artery	Mf	Multifidus muscle
CTA	Costotransverse articulation	LIPV	Left inferior pulmonary vein	PMa	Pectoralis major muscle
CVA	Costovertebral articulation	LLL	Left lower lobe of lung	PPl	Parietal pleura
Ca	Cartilage	LLLB	Left lower lobe bronchus	Pc	Pericardium
Es	Esophagus	LPA	Left pulmonary artery	PlPS	Pleuropericardial space
IMv	Internal mammary vessel	LV	Left ventricle of heart	PlS	Pleural space

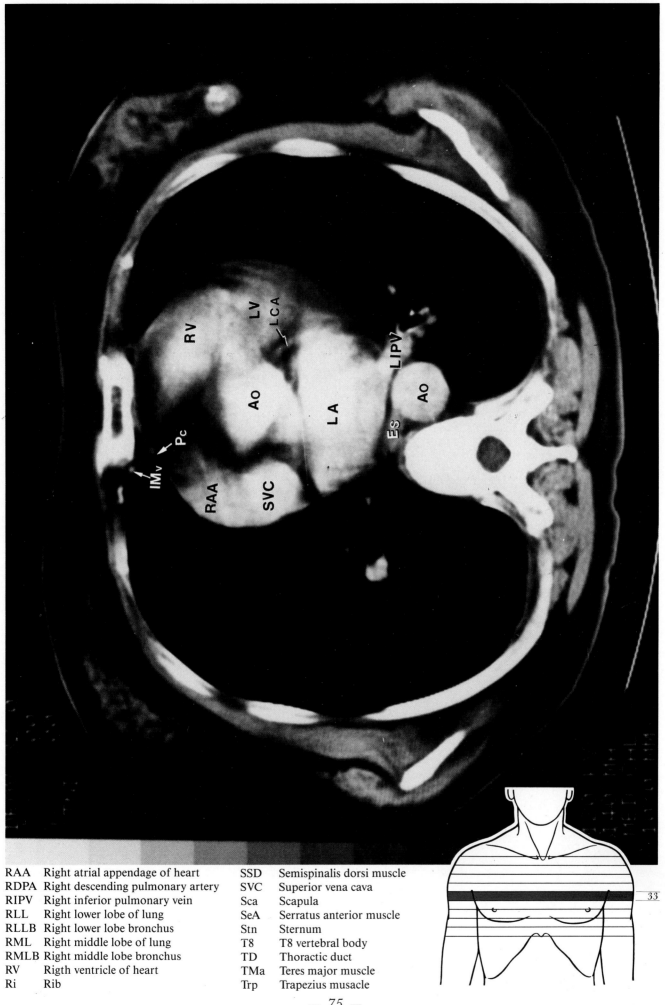

RAA	Right atrial appendage of heart	SSD	Semispinalis dorsi muscle
RDPA	Right descending pulmonary artery	SVC	Superior vena cava
RIPV	Right inferior pulmonary vein	Sca	Scapula
RLL	Right lower lobe of lung	SeA	Serratus anterior muscle
RLLB	Right lower lobe bronchus	Stn	Sternum
RML	Right middle lobe of lung	T8	T8 vertebral body
RMLB	Right middle lobe bronchus	TD	Thoractic duct
RV	Rigth ventricle of heart	TMa	Teres major muscle
Ri	Rib	Trp	Trapezius musacle

Plate 34. Transverse CHEST

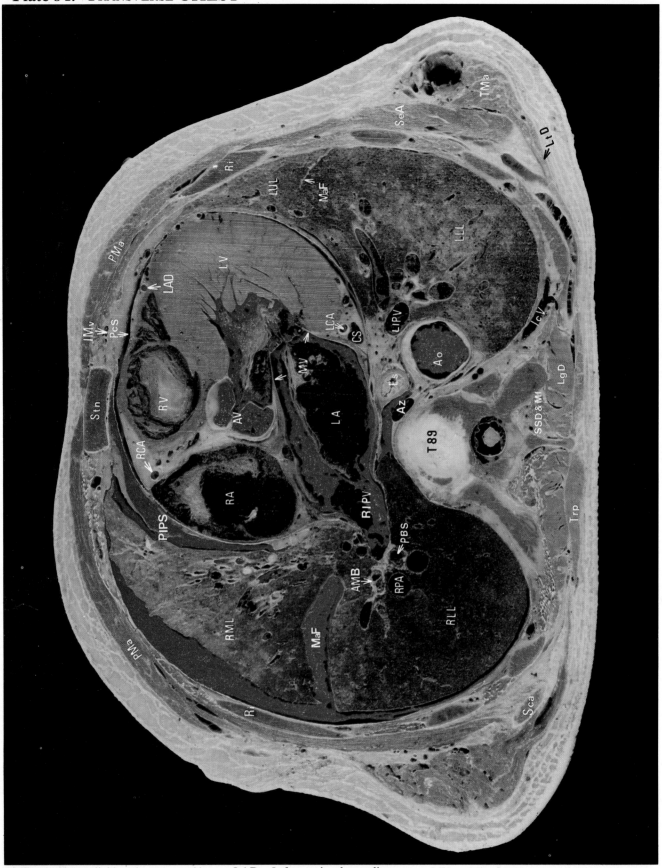

AMB	Anteromedial basal segment of lung	LAD	Left anterior descending branch of left coronary artery	MV	Mitral valve
AV	Aortic valve			MaF	Major fissure
Ao	Aorta	LCA	Left coronary artery	Mf	Multifidus muscle
Az	Azygos vein	LIPV	Left inferior pulmonary vein	PBS	Posterolateral basal segment of lung
CS	Coronary sinus	LLL	left lower lobe of lung	PMa	Pectoralis major muscle
Es	Esophagus	LUL	Left upper lobe of lung	PcS	Pericardial space
IMv	Internal mammary vessel	LV	Left ventricle of heart	PlPS	Pleuropericardial space
IcV	Intercostal vein	LgD	Longissimus dorsi muscle	RA	Right atrium of heart
LA	Left atrium of heart	LtD	Latissimus dorsi muscle	RCA	Right coronary artery

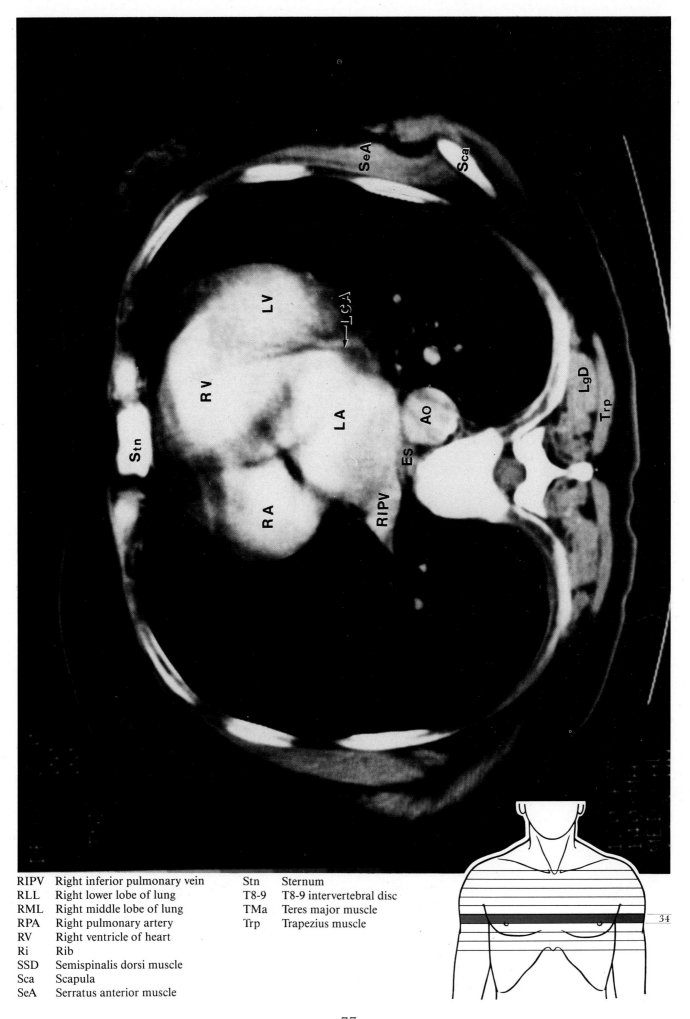

RIPV	Right inferior pulmonary vein	Stn	Sternum
RLL	Right lower lobe of lung	T8-9	T8-9 intervertebral disc
RML	Right middle lobe of lung	TMa	Teres major muscle
RPA	Right pulmonary artery	Trp	Trapezius muscle
RV	Right ventricle of heart		
Ri	Rib		
SSD	Semispinalis dorsi muscle		
Sca	Scapula		
SeA	Serratus anterior muscle		

Plate 35. TRANSVERSE CHEST

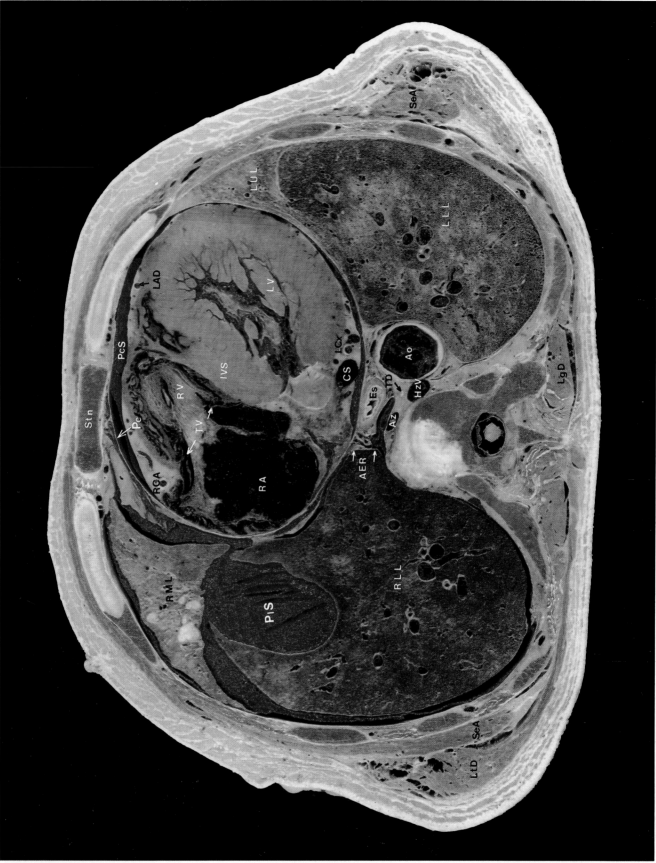

AER	Azygoesophageal recess	LCx	Left circumflex artery	RA	Right atrium of heart
Ao	Aorta	LLL	Left lower lobe of lung	RCA	Right coronary artery
Az	Azygos vein	LUL	Left upper lobe of lung	RLL	Right lower lobe of lung
CS	Coronary sinus	LV	Left ventricle of heart	RML	Right middle lobe of lung
Es	Esophagus	LgD	Longissimus dorsi muscle	RV	Right ventricle of heart
HzV	Hemiazygos vein	LtD	Latissimus dorsi muscle	SeA	Serratus anterior muscle
IVS	Interventricular septum	Pc	Pericardium	Stn	Sternum
LAD	Left anterior descending branch	PcS	Pericardial space	TD	Thoracic duct
	of left coronary artery	PlS	Pleural space	TV	Tricuspid valve

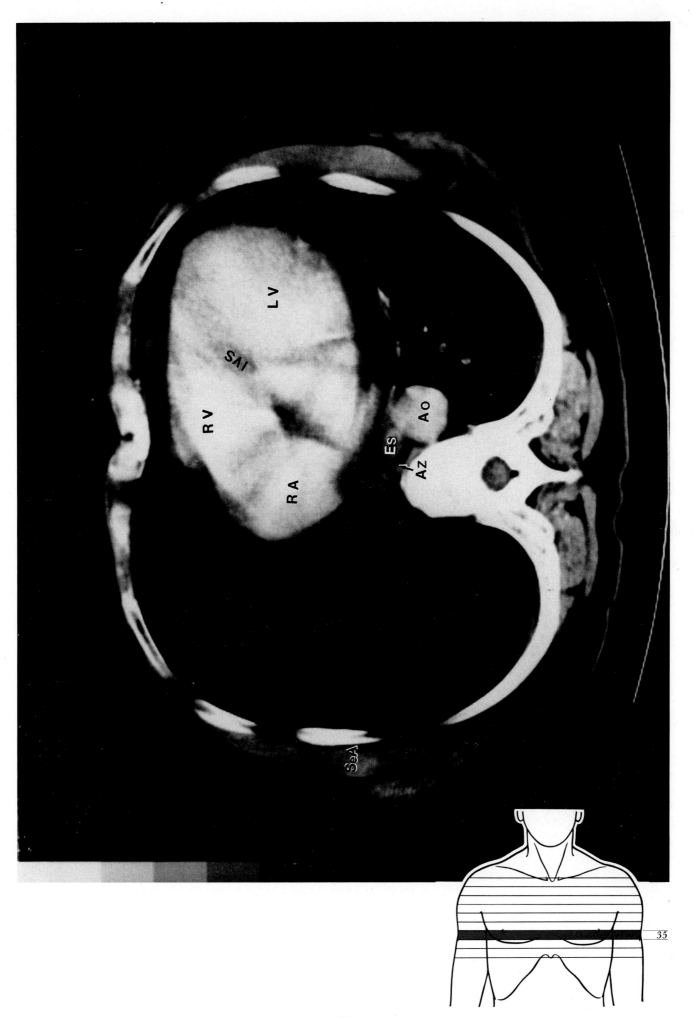

Plate 36. TRANSVERSE CHEST

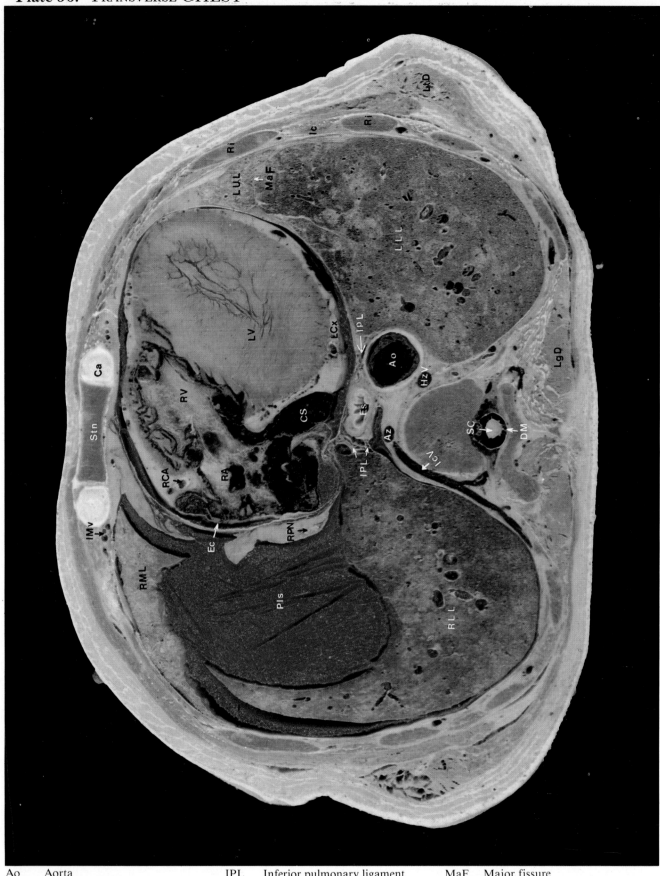

Ao	Aorta	IPL	Inferior pulmonary ligament	MaF	Major fissure
Az	Azygos vein	Ic	Intercostal muscle	PlS	Pleural space
CS	Coronary sinus	IcV	Intercostal vein	RA	Right atrium of heart
Ca	Cartilage	LCx	Left circumflex artery	RCA	Right coronary artery
DM	Dura mater	LLL	Left lower lobe of lung	RLL	Right lower lobe of lung
Ec	Epicardium	LUL	Left upper lobe of lung	RML	Right middle lobe of lung
Es	Esophagus	LV	Left venricle of heart	RPN	Right phrenic nerve
HzV	Hemiazygos vein	LgD	Longissimus dorsi muscle	RV	Right ventricle of heart
IMv	Internal mammary vessel	LtD	Latissimus dorsi muscle	Ri	Rib

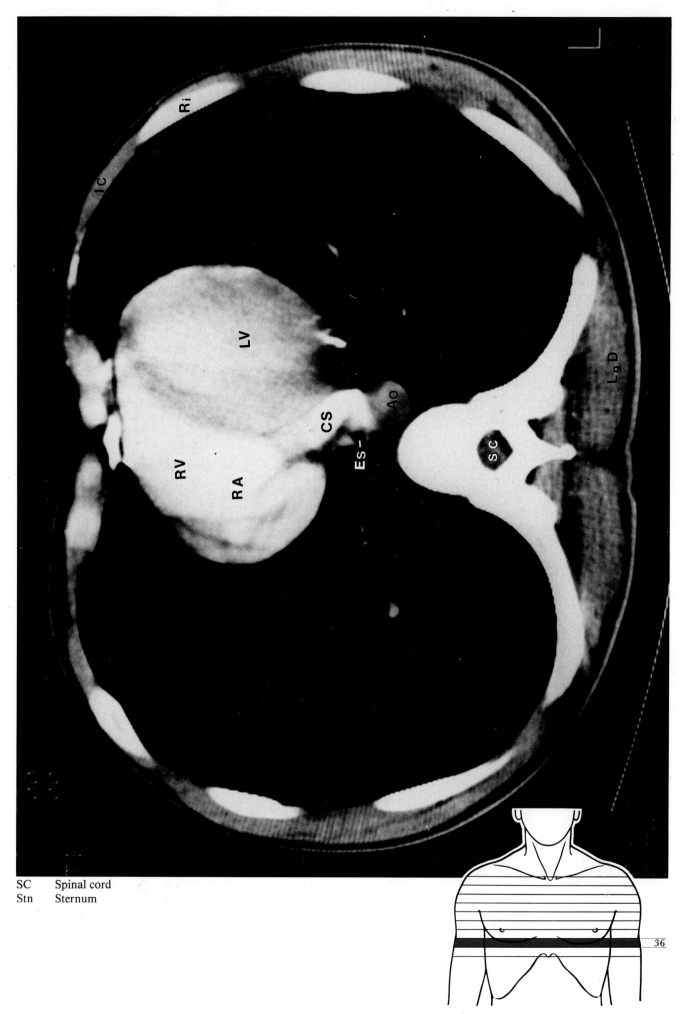

SC   Spinal cord
Stn   Sternum

Plate 37. TRANSVERSE CHEST

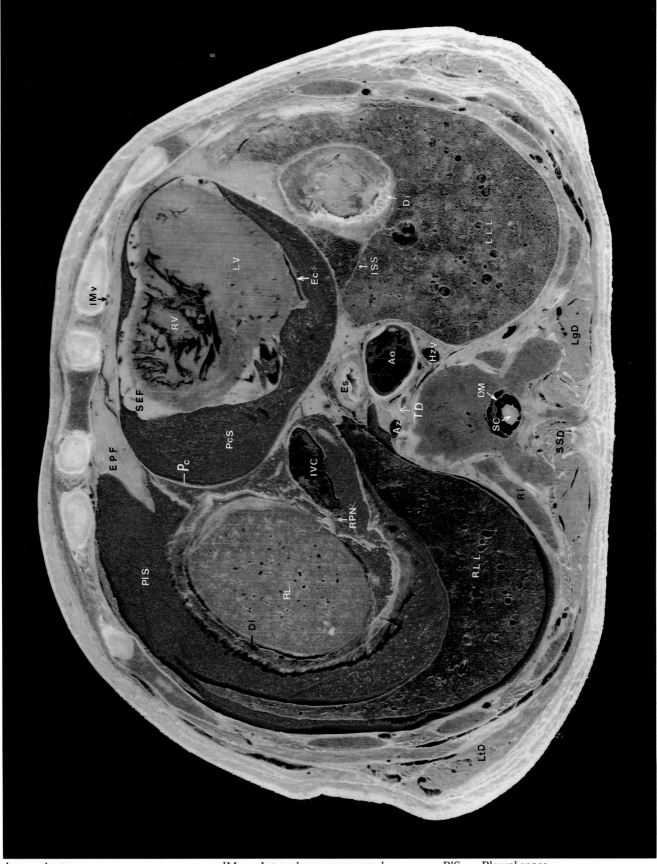

Ao	Aorta	IMv	Internal mammary vessel	PlS	Pleural space
Az	Azygos vein	ISS	Intersublobar septum	RL	Right lobe of liver
CS	Coronary sinus	IVC	Inferior vena cava	RLL	Right lower lobe of lung
DM	Dura mater	LLL	Left lower lobe of lung	RPN	Right phrenic nerve
Di	Diaphragm	LV	Left ventricle of heart	RV	Right ventricle of heart
EPF	Extrapericardial fat	LgD	Longissimus dorsi muscle	Ri	Rib
Ec	Epicardium	LtD	Latissimus dorsi muscle	SC	Spinal cord
Es	Esophagus	Pc	Pericardium	SEF	Subepicardial fat
HzV	Hemiazygos vein	PcS	Pericardial space	SSD	Semispinalis dorsi muscle

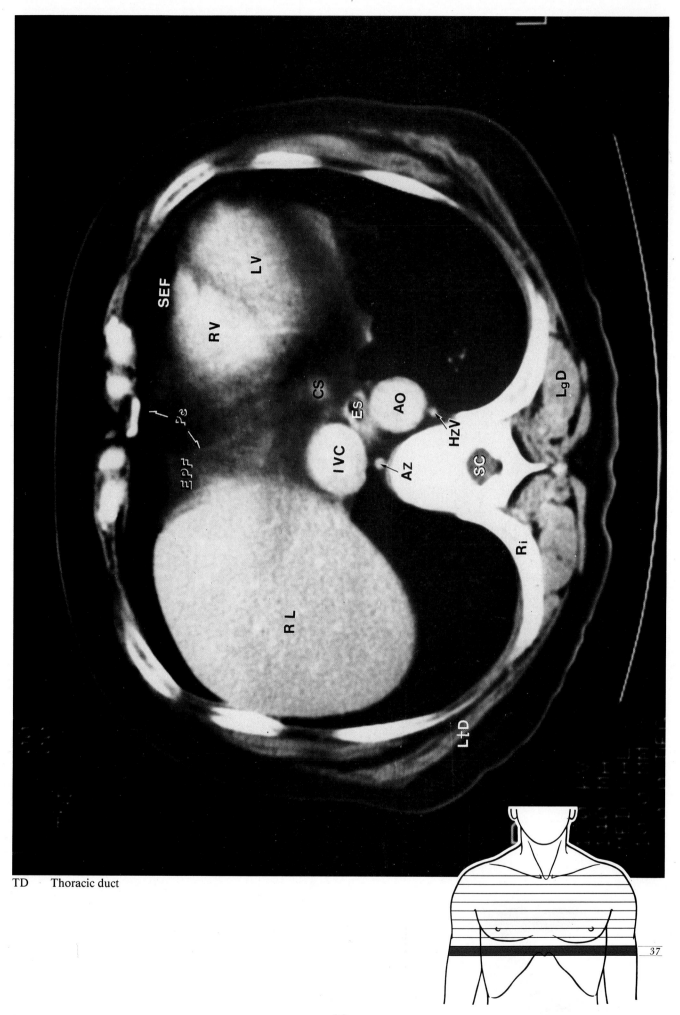

TD   Thoracic duct

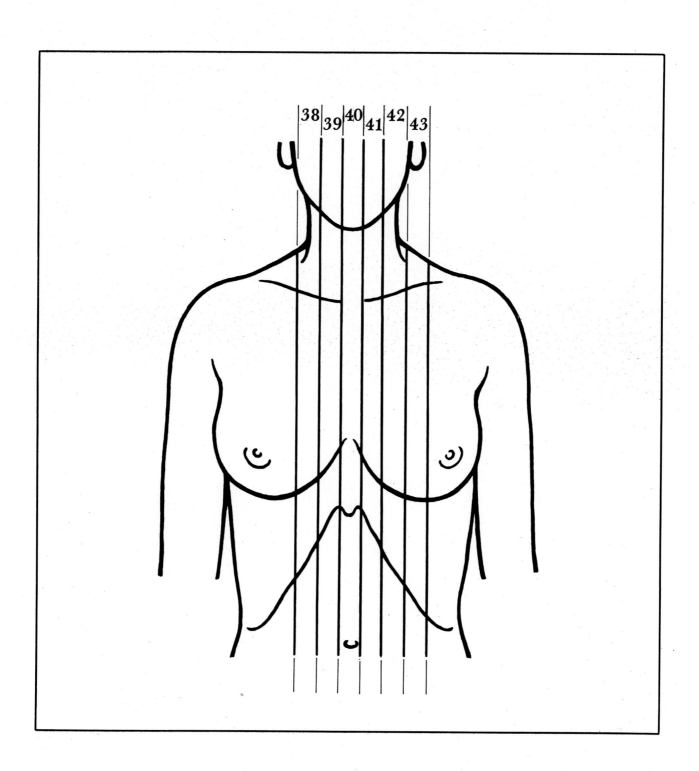

Sagittal Section-female (Specimen) Plate 38-43

Plate 38. SAGITTAL CHEST

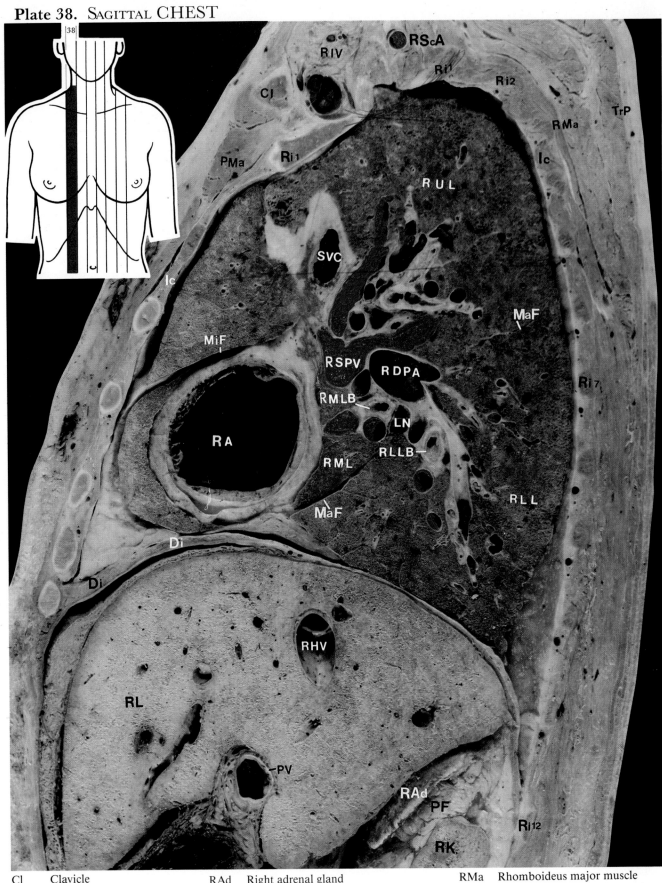

Cl	Clavicle	RAd	Right adrenal gland	RMa	Rhomboideus major muscle
Di	Diaphragm	RDPA	Right descending pulmonary artery	RSPV	Right superior pulmonary vein
Ic	Intercostal muscle	RHV	Right hepatic vein	RScA	Right subclavian artery
LN	Lymph node	RIV	Right innominate vein	RUL	Right upper lobe of lung
MaF	Major fissure	RK	Right kidney	Ri1	1st rib
MiF	Minor fissure	RL	Right lobe of liver	Ri2	2nd rib
PF	Perirenal fat	RLL	Right lower lobe of lung	Ri7	7th rib
PMa	Pectoralis major muscle	RLLB	Right lower lobe bronchus	Ri12	12th rib
PV	Portal vein	RML	Right middle lobe of lung	SVC	Superior vena cava
RA	Right atrium of heart	RMLB	Right middle lobe bronchus	TrP	Trapezius muscle

Plate 39. Sᴀɢɪᴛᴛᴀʟ CHEST

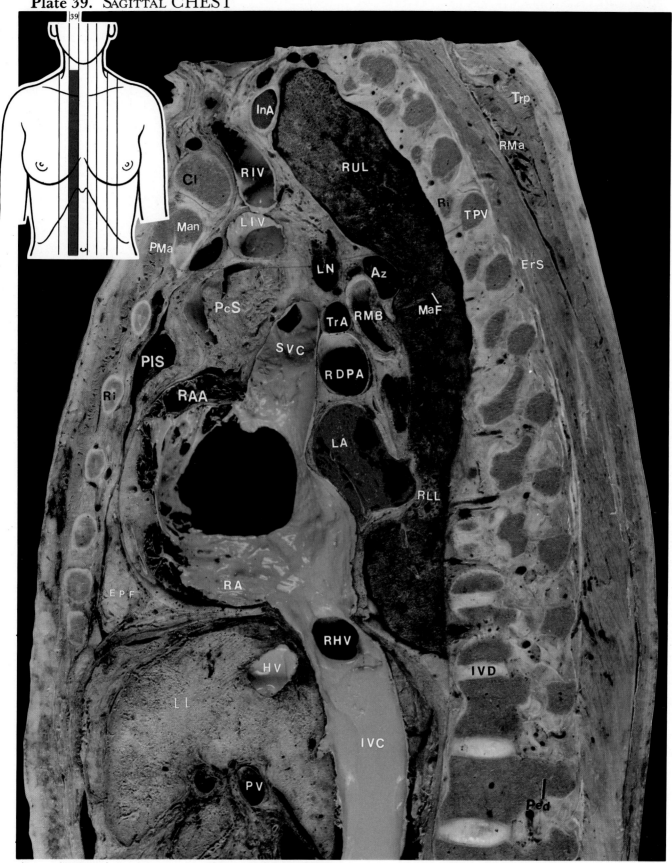

Az	Azygos vein	LL	Left lobe of liver	RDPA	Right descending pulmonary artery
Cl	Clavicle	LN	Lymph node	RHV	Right hepatic vein
EPF	Extrapericardial fat	Man	Manubrium	RIV	Right innominate vein
ErS	Erector spinae muscle	PMa	Pectoralis major muscle	RMB	Right main bronchus
HV	Hepatic vein	PV	Portal vein	RMa	Rhomboideus major muscle
IVC	Inferior vena cava	PcS	Pericardial space	RUL	Right upper lobe of lung
IVD	Intervertebral disc	Ped	Pedicle	Ri	Rib
InA	Innominate artery	PlS	Pleural space	SVC	Superior vena cava
LA	Left atrium of heart	RA	Right atrium of heart	TPV	Transverse process of vertebra
LIV	Left innominate vein	RAA	Right atrial appendage of heart	TrA	Truncus anterior

Plate 40. SAGITTAL CHEST

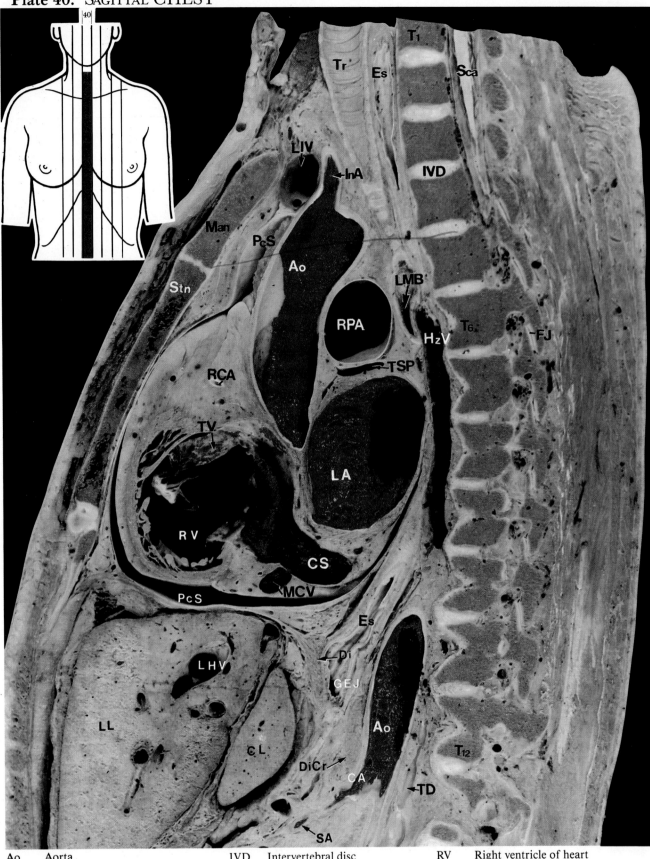

Ao	Aorta	IVD	Intervertebral disc	RV	Right ventricle of heart
CA	Celiac axis	InA	Innominate artery	SA	Splenic artery
CL	Caudate lobe of liver	LA	Left atrium of heart	Sca	Scapula
CS	Coronary sinus	LIV	Left inominate vein	Stn	Sternum
Di	Diaphragm	LMB	Left main bronchus	T1	T1 vertebral body
DiCr	Diaphragmatic crus	MCV	Middle cardiac vein	T6	T6 vertebral body
Es	Esophagus	Man	Manubrium	TD	Thoracic duct
FJ	Facet joint of thoracic spine	PcS	Pericardial space	TSP	Transverse sinus of pericardium
GEJ	Gastroesophageal junction	RCA	Right coronary artery	TV	Tricuspid valve
HzV	Hemiazygos vein	RPA	Right pulmonary artery	Tr	Trachea

Plate 41. Sᴀɢɪᴛᴛᴀʟ CHEST

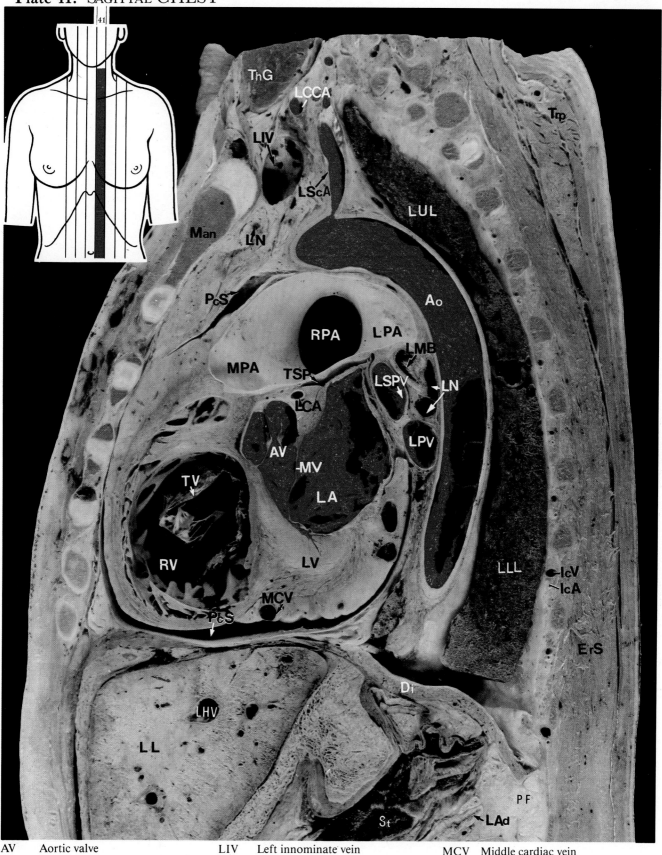

AV	Aortic valve		LIV	Left innominate vein		MCV	Middle cardiac vein
Ao	Aorta		LL	Left lobe of liver		MPA	Main pulmonary artery
Di	Diaphragm		LLL	Left lower lobe of lung		MV	Mitral valve
ErS	Erector spinae muscle		LMB	Left main bronchus		PF	Perirenal fat
IcA	Intercostal artery		LN	Lymph node		RPA	Right pulmonary artery
IcV	Intercostal vein		LPA	Left pulmonary artery		RV	Right ventricle of heart
LA	Left atrium of heart		LPV	Left portal vein		St	Stomach
LAd	Left adrenal gland		LSPV	Left superior pulmonary vein		TSP	Transverse sinus of pericardium
LCA	Left coronary artery		LScA	Left subclavian artery		TV	Tricuspid valve
LCCA	Left common carotid artery		LUL	Left upper lobe of lung		ThG	Thyroid gland
LHV	Left hepatic vein		LV	Left ventricle of heart		Trp	Trapezius muscle

Plate 42. SAGITTAL CHEST

Cl	Clavicle	LPA	Left pulmonary artery	Pap	Papillary muscle
Di	Diaphragm	LSPV	Left superior pulmonary vein	PcS	Pericardial space
GHR	Gastrohepatic recess	LScA	Left subclavian artery	PlS	Pleural space
LAA	Left atrial appendage of heart	LUL	Left upper lobe of lung	RV	Right ventricle of heart
LCA	Left coronary artery	LULB	Left upper lobe bronchus	Ri1	1st rib
LIV	Left innominate vein	LV	Left ventricle of heart	Spl	Spleen
LL	Left lobe of liver	MCV	Middle cardiac vein	St	Stomach
LLL	Left lower lobe of lung	MaF	Major fissure	Trp	Trapezius muscle
LLLB	Left lower lobe bronchus	PMa	Pectoralis major muscle		
LN	Lymph node	PV	Pulmonic valve		

Plate 43. SAGITTAL CHEST

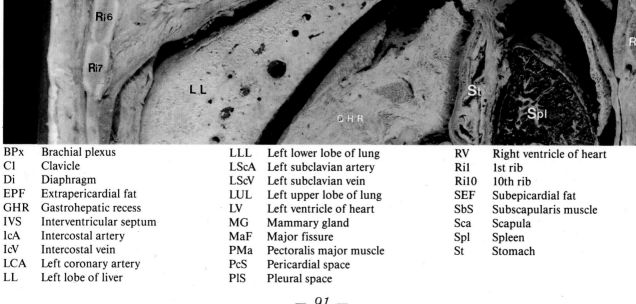

BPx	Brachial plexus	LLL	Left lower lobe of lung	RV	Right ventricle of heart
Cl	Clavicle	LScA	Left subclavian artery	Ri1	1st rib
Di	Diaphragm	LScV	Left subclavian vein	Ri10	10th rib
EPF	Extrapericardial fat	LUL	Left upper lobe of lung	SEF	Subepicardial fat
GHR	Gastrohepatic recess	LV	Left ventricle of heart	SbS	Subscapularis muscle
IVS	Interventricular septum	MG	Mammary gland	Sca	Scapula
IcA	Intercostal artery	MaF	Major fissure	Spl	Spleen
IcV	Intercostal vein	PMa	Pectoralis major muscle	St	Stomach
LCA	Left coronary artery	PcS	Pericardial space		
LL	Left lobe of liver	PlS	Pleural space		

PART 3
ABDOMEN AND PELVIS

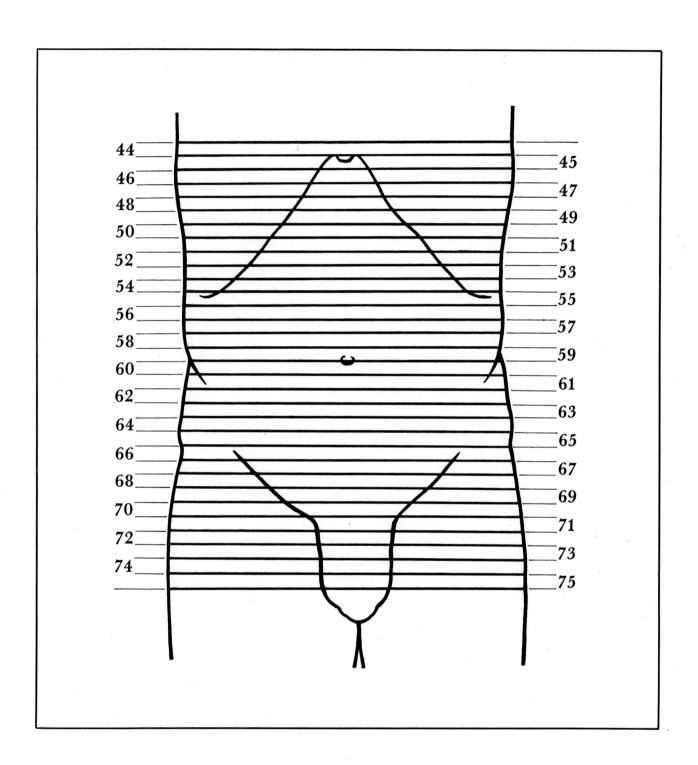

Transverse Section-male (Specimen & CT) Plate 44-75

Plate 44. TRANSVERSE ABDOMEN

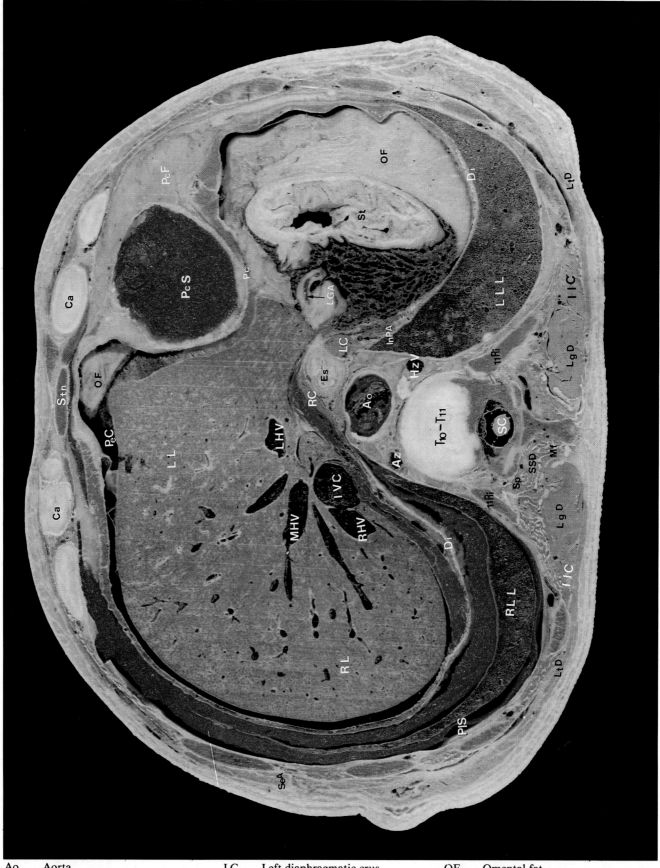

Ao	Aorta	LC	Left diaphragmatic crus	OF	Omental fat
Az	Azygos vein	LGA	Left gastric artery	Pc	Pericardium
Ca	Cartilage	LHV	Left hepatic vein	PcF	Pericardial fat
Di	Diaphragm	LL	Left lobe of liver	PcS	Pericardial space
Es	Esophagus	LLL	Left lower lobe of lung	PeC	Peritoneal cavity
HzV	Hemiazygos vein	LgD	Longissimus dorsi muscle	PlS	Pleural space
IVC	Inferior vena cava	LtD	Latissimus dorsi muscle	RC	Right diaphragmatic crus
IIC	Iliocostalis muscle	MHV	Middle hepatic vein	RHV	Right hepatic vein
InPA	Inferior phrenic artery	Mf	Multifidus muscle	RL	Right lobe of liver

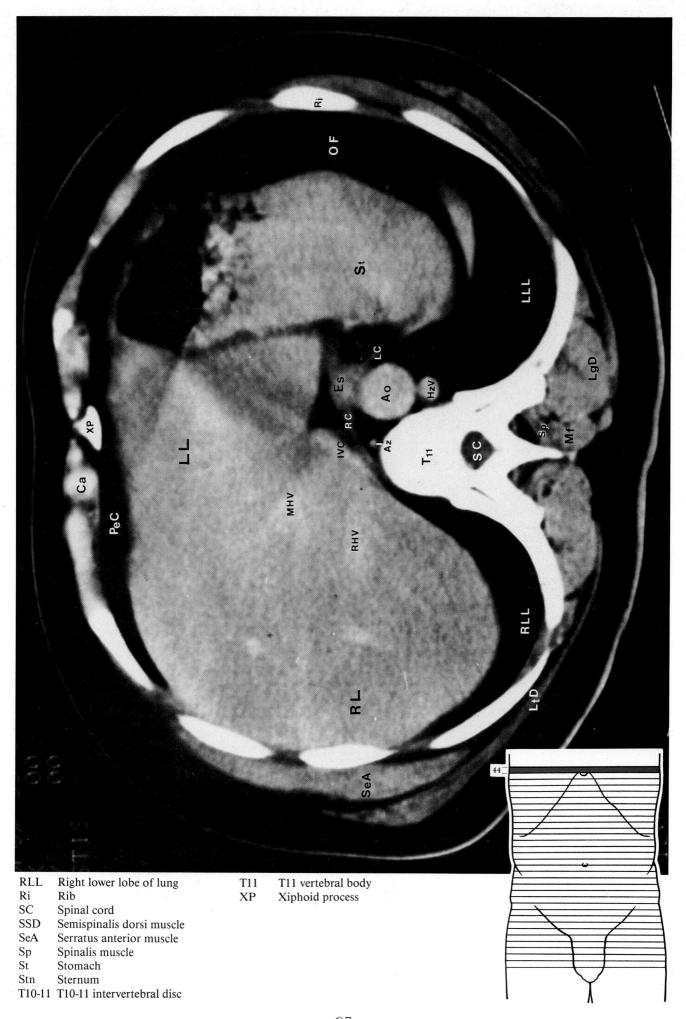

RLL	Right lower lobe of lung		T11	T11 vertebral body
Ri	Rib		XP	Xiphoid process
SC	Spinal cord			
SSD	Semispinalis dorsi muscle			
SeA	Serratus anterior muscle			
Sp	Spinalis muscle			
St	Stomach			
Stn	Sternum			
T10-11	T10-11 intervertebral disc			

Plate 45. TRANSVERSE ABDOMEN

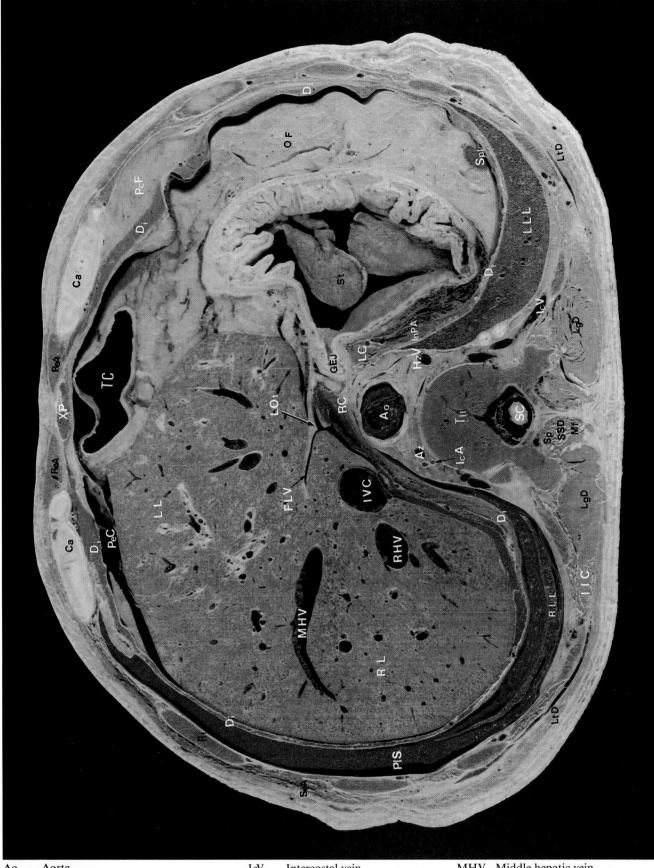

Ao	Aorta	IcV	Intercostal vein	MHV	Middle hepatic vein
Az	Azygos vein	IlC	Iliocostalis muscle	Mf	Multifidus muscle
Ca	Cartilage	InPA	Inferior phrenic artery	OF	Omental fat
Di	Diaphragm	LC	Left diaphragmatic crus	PcF	Pericardial fat
FLV	Fissure for ligamentum venosum	LL	Left lobe of liver	PeC	Peritoneal cavity
GEJ	Gastroesophageal junction	LLL	Left lower lobe of lung	PlS	Pleural space
HzV	Hemiazygos vein	LOt	Lesser omentum	RC	Right diaphragmatic crus
IVC	Inferior vena cava	LgD	Longissimus dorsi muscle	RHV	Right hepatic vein
IcA	Intercostal artery	LtD	Latissimus dorsi muscle	RL	Right lobe of liver

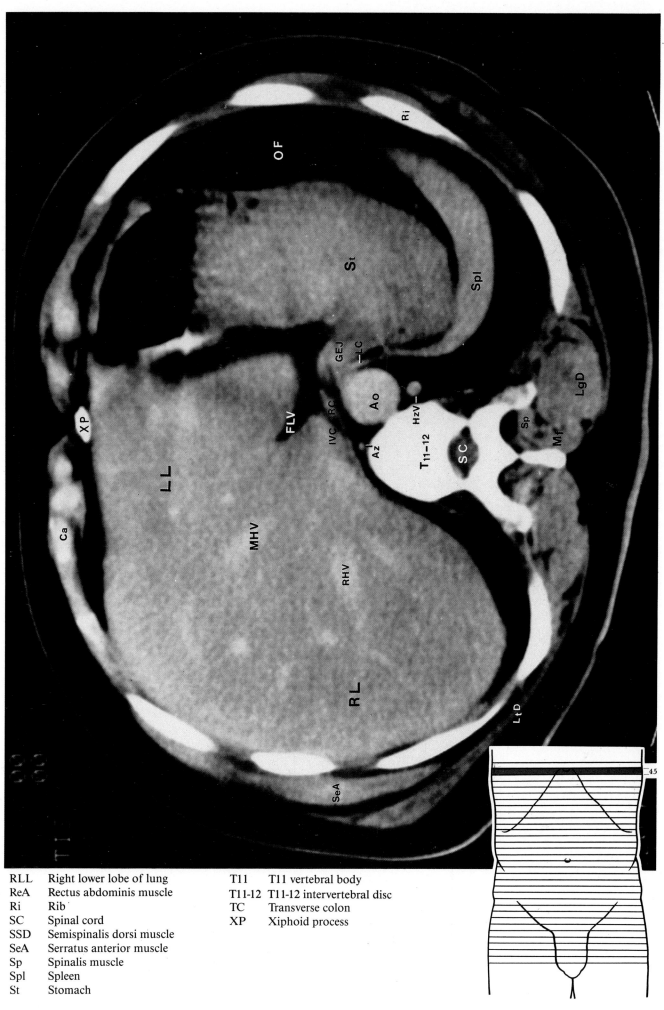

RLL	Right lower lobe of lung
ReA	Rectus abdominis muscle
Ri	Rib
SC	Spinal cord
SSD	Semispinalis dorsi muscle
SeA	Serratus anterior muscle
Sp	Spinalis muscle
Spl	Spleen
St	Stomach

T11	T11 vertebral body
T11-12	T11-12 intervertebral disc
TC	Transverse colon
XP	Xiphoid process

Plate 46. TRANSVERSE ABDOMEN

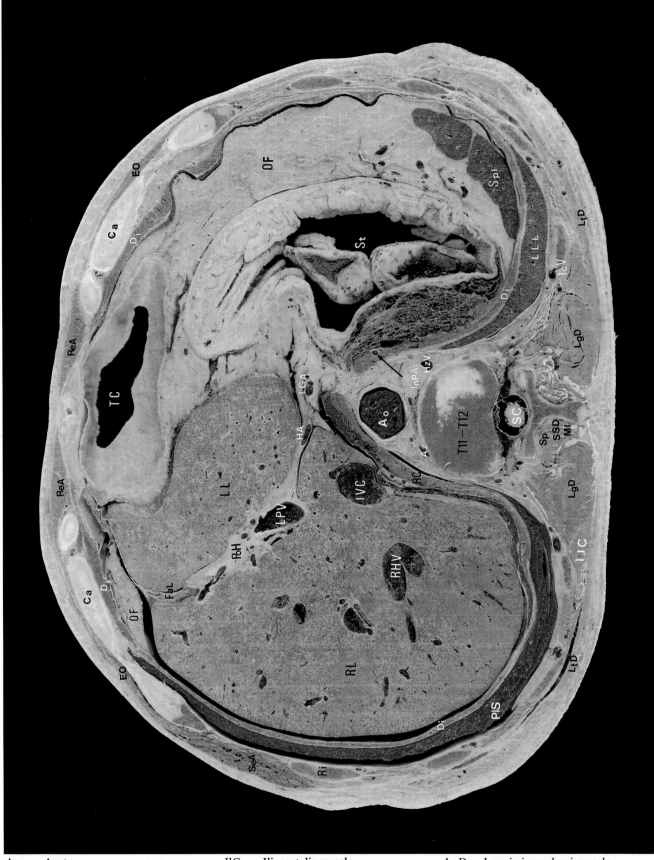

Ao	Aorta	IlC	Iliocostalis muscle	LgD	Longissimus dorsi muscle
Az	Azygos vein	IcA	Intercostal artery	LtD	Latissimus dorsi muscle
Ca	Cartilage	IcV	Intercostal vein	Mf	Multifidus muscle
Di	Diaphragm	InPA	Inferior phrenic artery	OF	Omental fat
EO	External oblique muscle	LC	Left diaphragmatic crus	PlS	Pleural space
FaL	Falciform ligament	LGA	Left gastric artery	PoH	Porta hepatis
HA	Hepatic artery	LL	Left lobe of liver	RC	Right diaphragmatic crus
HzV	Hemiazygos vein	LLL	Left lower lobe of lung	RHV	Right hepatic vein
IVC	Inferior vena cava	LPV	Left portal vein	RL	Right lobe of liver

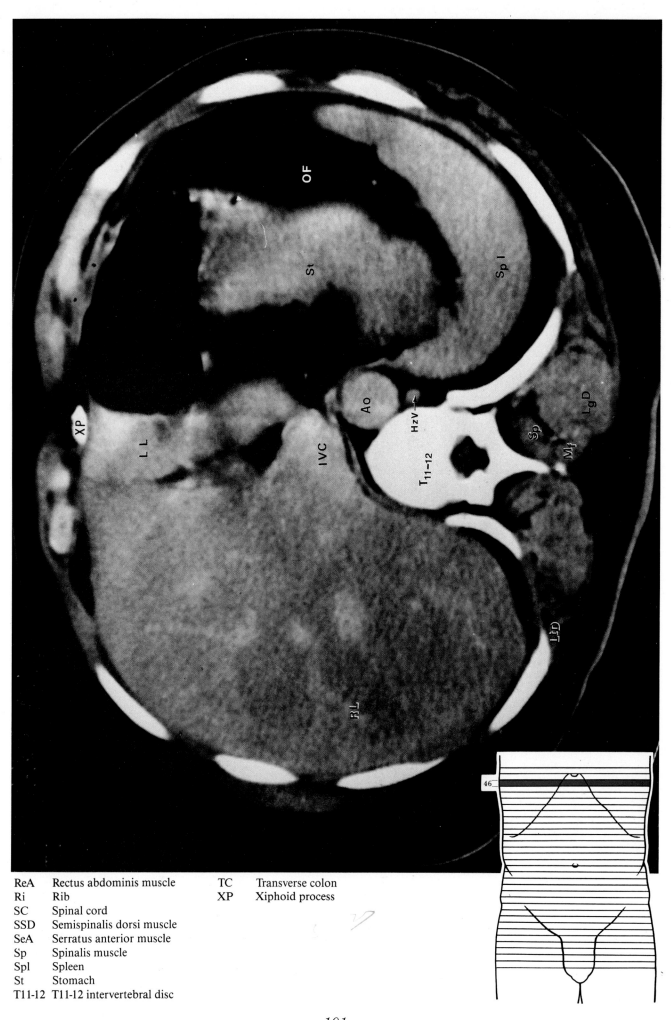

ReA	Rectus abdominis muscle
Ri	Rib
SC	Spinal cord
SSD	Semispinalis dorsi muscle
SeA	Serratus anterior muscle
Sp	Spinalis muscle
Spl	Spleen
St	Stomach
T11-12	T11-12 intervertebral disc

TC	Transverse colon
XP	Xiphoid process

Plate 47. TRANSVERSE ABDOMEN

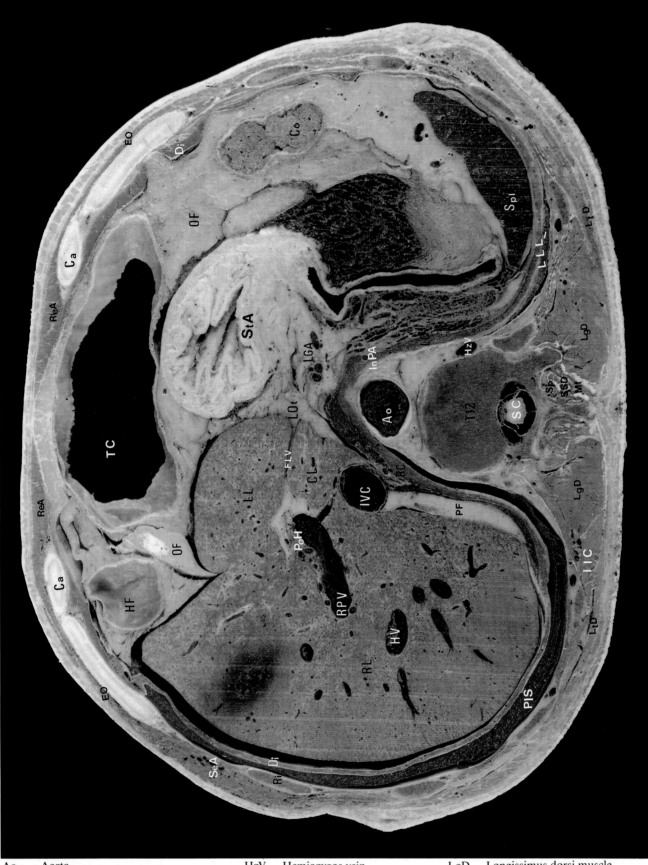

Ao	Aorta	HzV	Hemiazygos vein	LgD	Longissimus dorsi muscle
CL	Caudate lobe of liver	IVC	Inferior vena cava	LtD	Latissimus dorsi muscle
Ca	Cartilage	IlC	Iliocostalis muscle	Mf	Multifidus muscle
Co	Colon	InPA	Inferior phrenic artery	OF	Omental fat
Di	Diaphragm	LC	Left diaphragmatic crus	PB	Pancreas, body
EO	External oblique muscle	LGA	Left gastric artery	PF	Perirenal fat
FLV	Fissure for ligamentum venosum	LL	Left lobe of liver	PT	Pancreas, tail
HF	Hepatic flexure of colon	LLL	Left lower lobe of lung	PV	Portal vein
HV	Hepatic vein	LOt	Lesser omentum	PlS	Pleural space

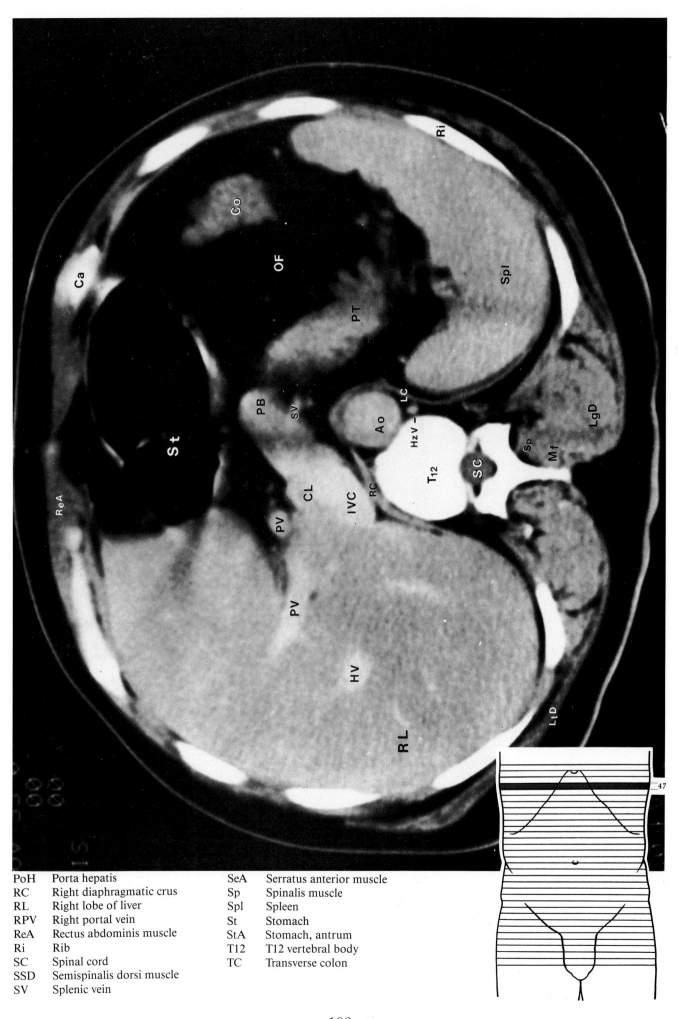

PoH	Porta hepatis	SeA	Serratus anterior muscle
RC	Right diaphragmatic crus	Sp	Spinalis muscle
RL	Right lobe of liver	Spl	Spleen
RPV	Right portal vein	St	Stomach
ReA	Rectus abdominis muscle	StA	Stomach, antrum
Ri	Rib	T12	T12 vertebral body
SC	Spinal cord	TC	Transverse colon
SSD	Semispinalis dorsi muscle		
SV	Splenic vein		

Plate 48. TRANSVERSE ABDOMEN

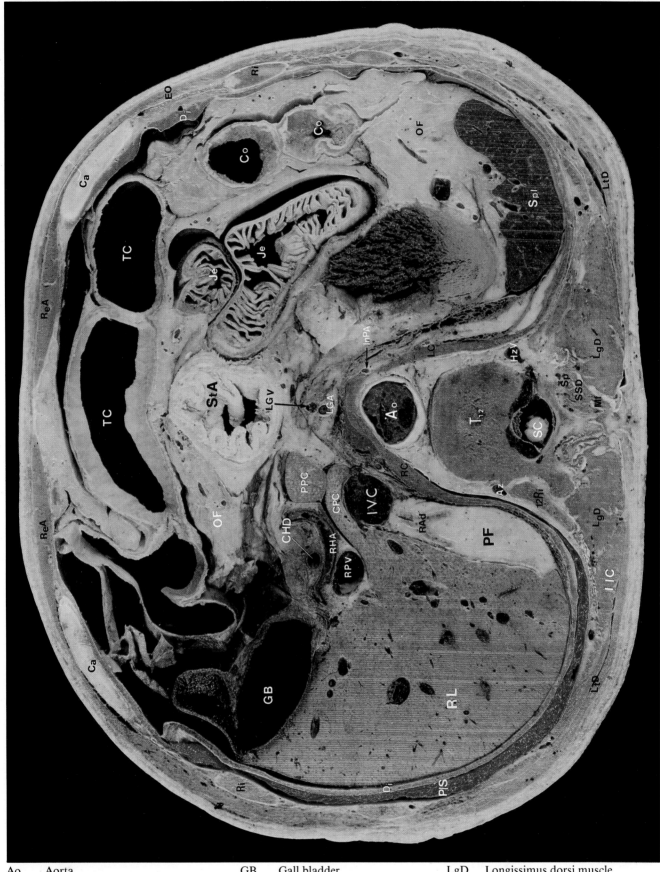

Ao	Aorta	GB	Gall bladder	LgD	Longissimus dorsi muscle
Az	Azygos vein	HzV	Hemiazygos vein	LtD	Latissimus dorsi muscle
CHD	Common hepatic duct	IVC	Inferior vena cava	MeF	Mesenteric fat
CPC	Caudate process of caudate lobe	IlC	Iliocostalis muscle	Mf	Multifidus muscle
Ca	Cartilage	InPA	Inferior phrenic artery	OF	Omental fat
Co	Colon	Je	Jejunum	PB	Pancreas, body
DB	Duodenal bulb	LC	Left diaphragmatic crus	PF	Perirenal fat
Di	Diaphragm	LGA	Left gastric artery	PPC	Papillary process of caudate lobe
EO	External oblique muscle	LGV	Left gastric vein	PT	Pancreas, tail

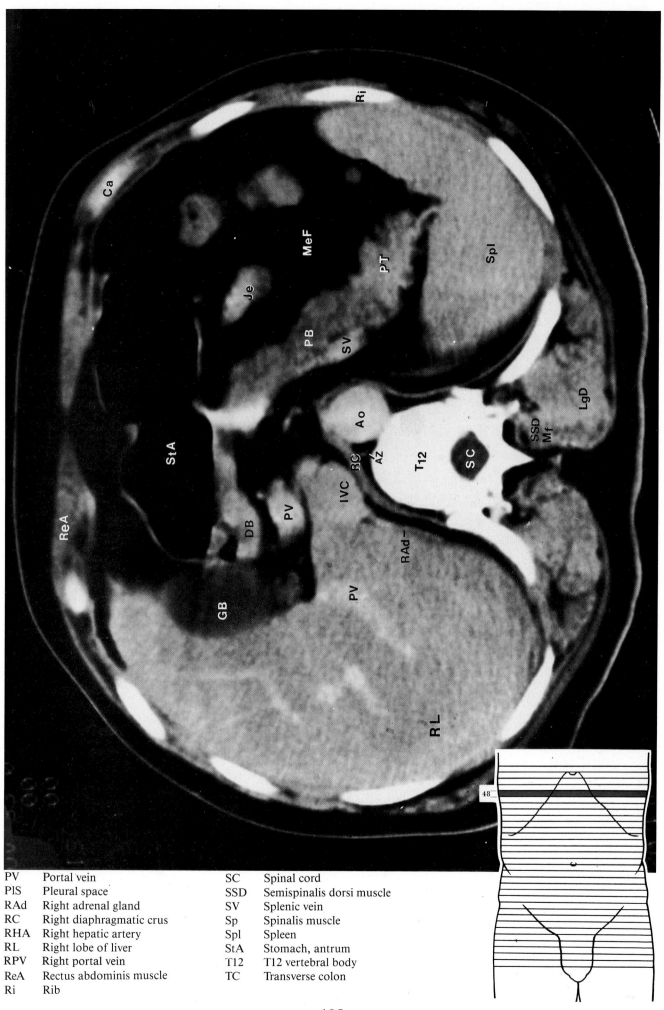

PV	Portal vein	SC	Spinal cord
PlS	Pleural space	SSD	Semispinalis dorsi muscle
RAd	Right adrenal gland	SV	Splenic vein
RC	Right diaphragmatic crus	Sp	Spinalis muscle
RHA	Right hepatic artery	Spl	Spleen
RL	Right lobe of liver	StA	Stomach, antrum
RPV	Right portal vein	T12	T12 vertebral body
ReA	Rectus abdominis muscle	TC	Transverse colon
Ri	Rib		

Plate 49. TRANSVERSE ABDOMEN

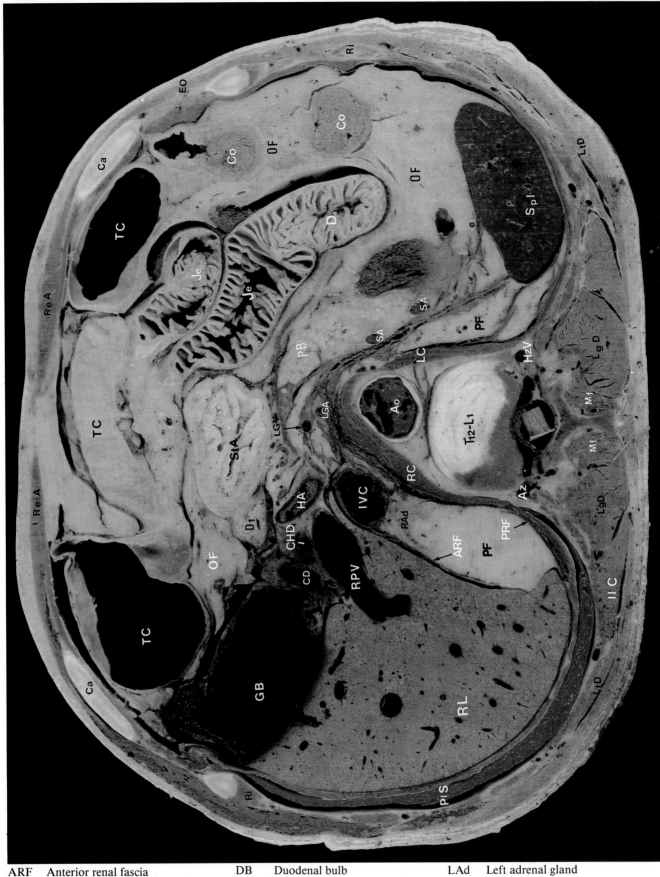

ARF	Anterior renal fascia	DB	Duodenal bulb	LAd	Left adrenal gland
Ao	Aorta	EO	External oblique muscle	LC	Left diaphragmatic crus
Az	Azygos vein	GB	Gall bladder	LGA	Left gastric artery
CD	Cystic duct	HA	Hepatic artery	LGV	Left gastric vein
CHD	Common hepatic duct	HzV	Hemiazygos vein	LK	Left kidney
Ca	Cartilage	IVC	Inferior vena cava	LgD	Longissimus dorsi muscle
Co	Colon	IlC	Iliocostalis muscle	LtD	Latissimus dorsi muscle
D1	First portion of duodenum	Je	Jejunum	Mf	Multifidus muscle
D4	Fourth portion of duodenum	L1	L1 vertebral body	OF	Omental fat

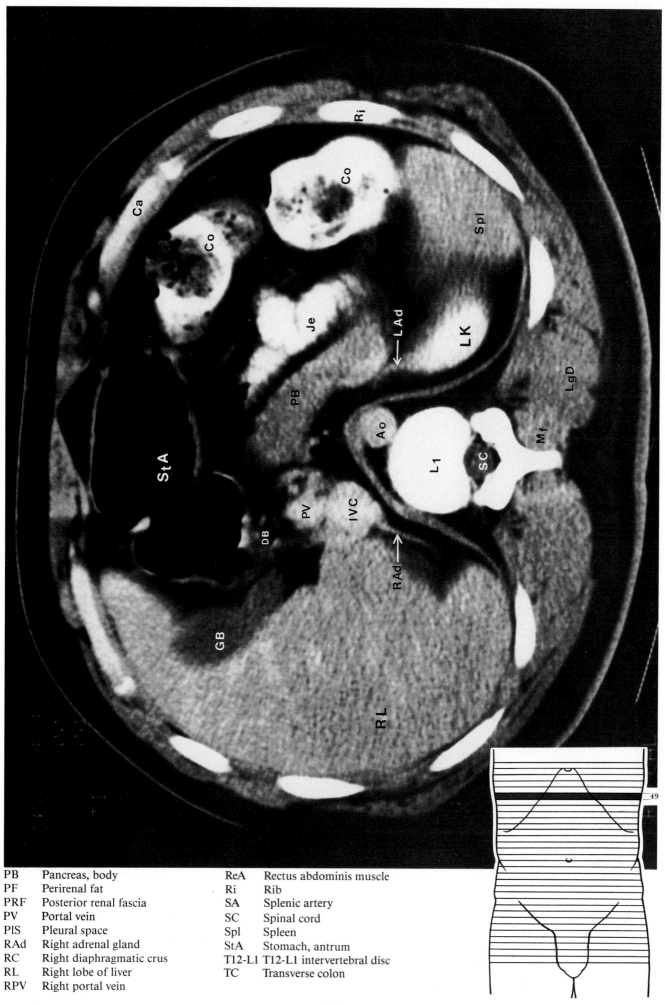

PB	Pancreas, body	ReA	Rectus abdominis muscle
PF	Perirenal fat	Ri	Rib
PRF	Posterior renal fascia	SA	Splenic artery
PV	Portal vein	SC	Spinal cord
PlS	Pleural space	Spl	Spleen
RAd	Right adrenal gland	StA	Stomach, antrum
RC	Right diaphragmatic crus	T12-L1	T12-L1 intervertebral disc
RL	Right lobe of liver	TC	Transverse colon
RPV	Right portal vein		

Plate 50. TRANSVERSE ABDOMEN

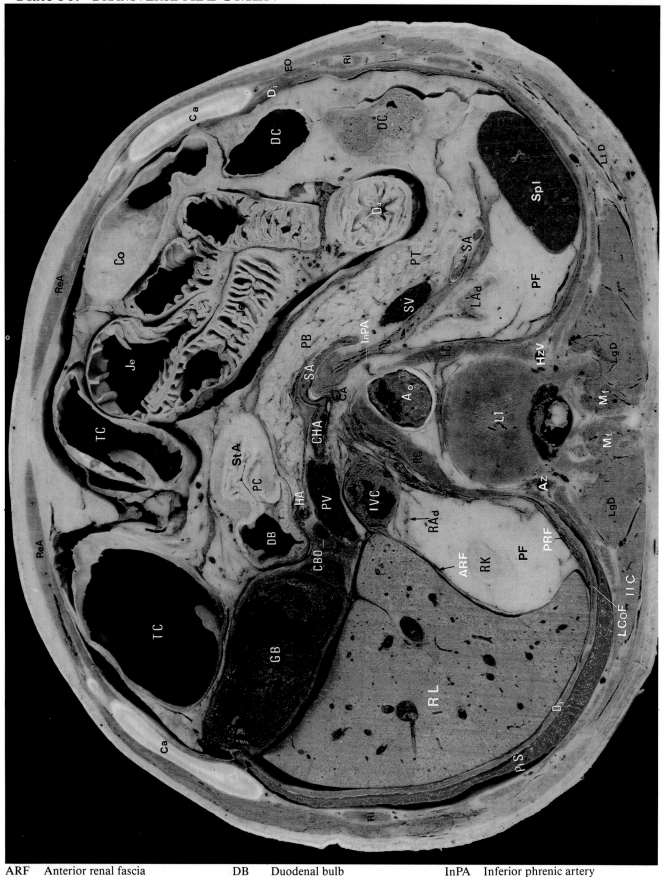

ARF	Anterior renal fascia	DB	Duodenal bulb	InPA	Inferior phrenic artery
Ao	Aorta	DC	Descending colon	Je	Jejunum
Az	Azygos vein	Di	Diaphragm	L1	L1 vertebral body
CA	Celiac axis	EO	External oblique muscle	LAd	Left adrenal gland
CBD	Common bile duct	GB	Gall bladder	LC	Left diaphragmatic crus
CHA	Common hepatic artery	HA	Hepatic artery	LCoF	Lateroconal fascia
Ca	Cartilage	HzV	Hemiazygos vein	LK	Left kidney
Co	Colon	IVC	Inferior vena cava	LL	Left lobe of liver
D4	Fourth portion of duodenum	IlC	Iliocostalis muscle	LgD	Longissimus dorsi muscle

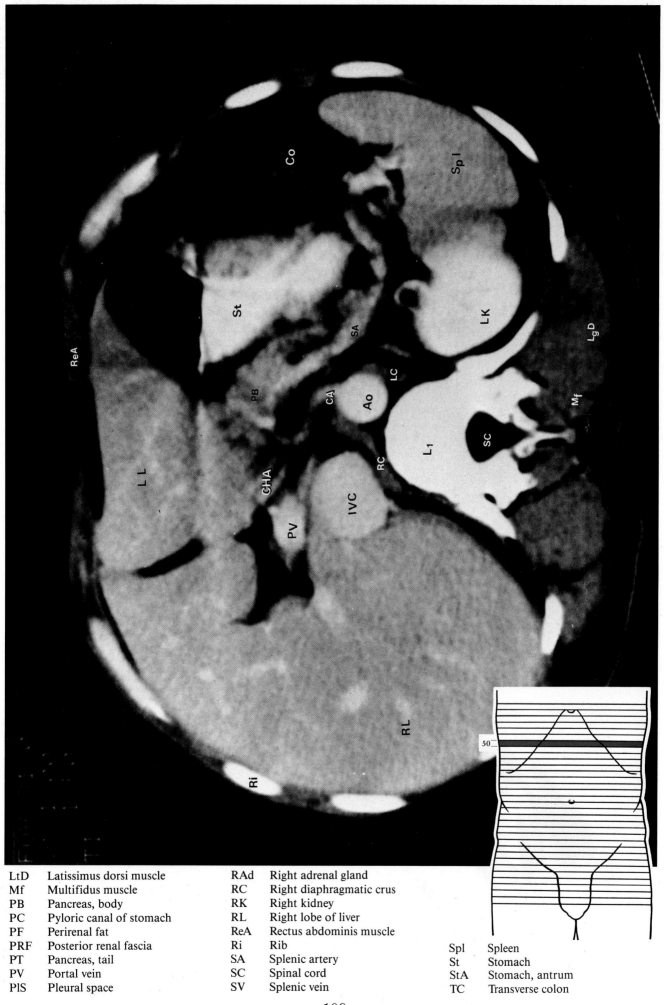

LtD	Latissimus dorsi muscle	RAd	Right adrenal gland	
Mf	Multifidus muscle	RC	Right diaphragmatic crus	
PB	Pancreas, body	RK	Right kidney	
PC	Pyloric canal of stomach	RL	Right lobe of liver	
PF	Perirenal fat	ReA	Rectus abdominis muscle	
PRF	Posterior renal fascia	Ri	Rib	
PT	Pancreas, tail	SA	Splenic artery	
PV	Portal vein	SC	Spinal cord	Spl	Spleen
PlS	Pleural space	SV	Splenic vein	St	Stomach
				StA	Stomach, antrum
				TC	Transverse colon

Plate 51. Tʀᴀɴꜱᴠᴇʀꜱᴇ ABDOMEN

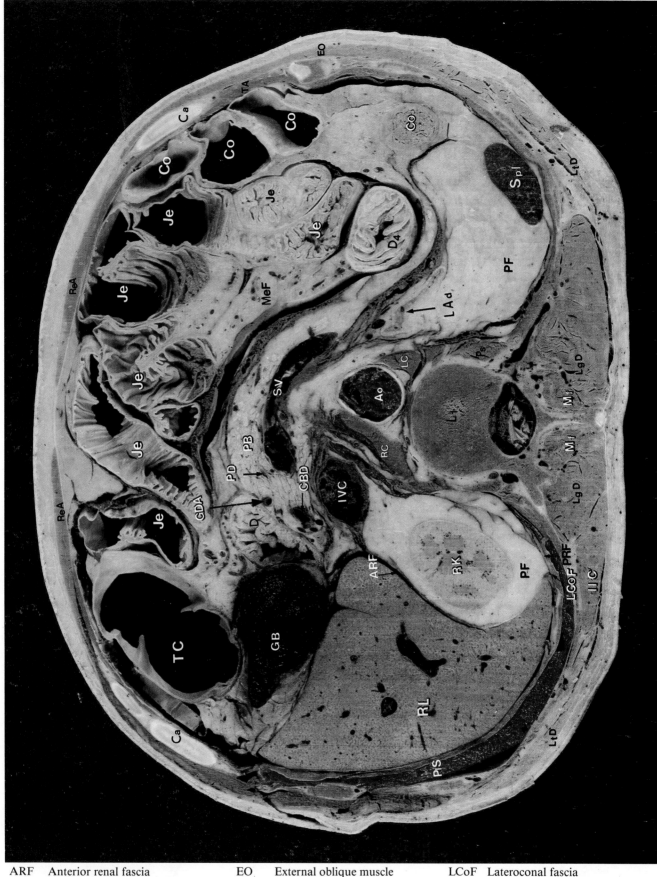

ARF	Anterior renal fascia	EO	External oblique muscle	LCoF	Lateroconal fascia
Ao	Aorta	GB	Gall bladder	LgD	Longissimus dorsi muscle
Az	Azygos vein	GDA	Gastroduodenal artery	LtD	Latissimus dorsi muscle
CBD	Common bile duct	IVC	Inferior vena cava	MeF	Mesenteric fat
Ca	Cartilage	IIC	Iliocostalis muscle	Mf	Multifidus muscle
Co	Colon	Je	Jejunum	PB	Pancreas, body
D1	First portion of duodenum	L1	L1 vertebral body	PD	Pancreatic duct
D4	Fourth portion of duodenum	LAd	Left adrenal gland	PF	Perirenal fat
DB	Duodenal bulb	LC	Left diaphragmatic crus	PRF	Posterior renal fascia

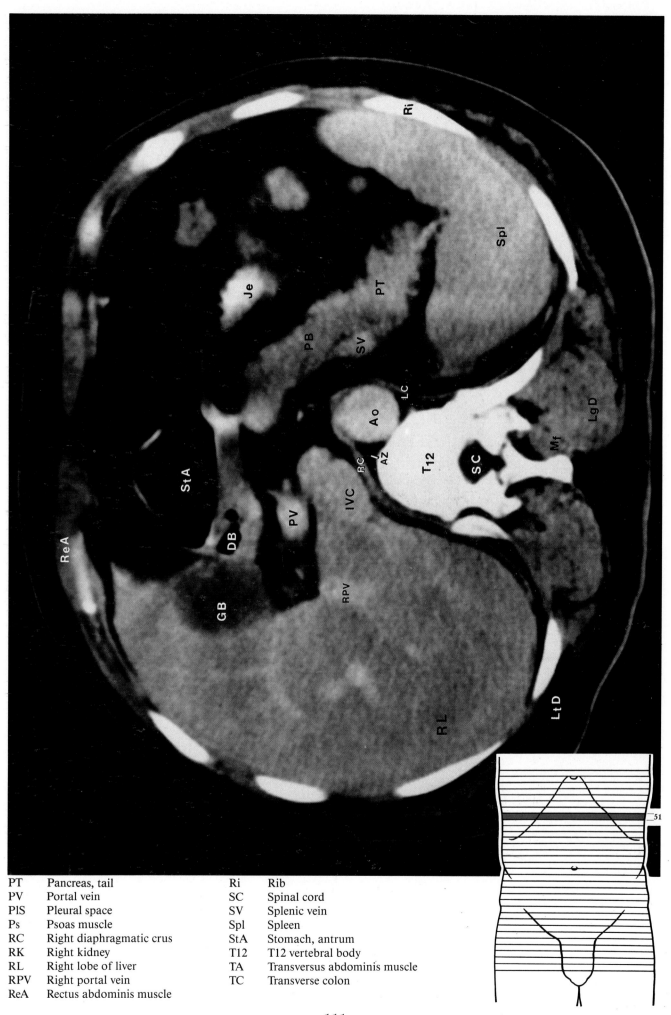

PT	Pancreas, tail	Ri	Rib
PV	Portal vein	SC	Spinal cord
PlS	Pleural space	SV	Splenic vein
Ps	Psoas muscle	Spl	Spleen
RC	Right diaphragmatic crus	StA	Stomach, antrum
RK	Right kidney	T12	T12 vertebral body
RL	Right lobe of liver	TA	Transversus abdominis muscle
RPV	Right portal vein	TC	Transverse colon
ReA	Rectus abdominis muscle		

Plate 52. TRANSVERSE ABDOMEN

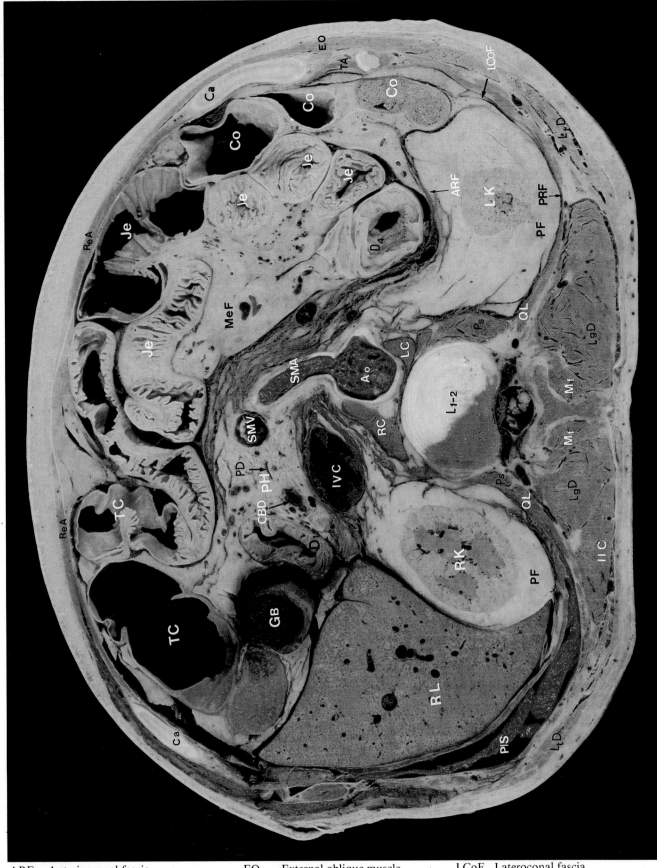

ARF	Anterior renal fascia	
Ao	Aorta	
Az	Azygos vein	
CBD	Common bile duct	
Ca	Cartilage	
Co	Colon	
D1	First portion of duodenum	
D2	Second portion of duodenum	
D4	Fourth portion of duodenum	

EO	External oblique muscle
GB	Gall bladder
HzV	Hemiazygos vein
IVC	Inferior vena cava
IlC	Iliocostalis muscle
Je	Jejunum
L1	L1 vertebral body
L1-2	L1-2 intervertebral disc
LC	Left diaphragmatic crus

LCoF	Lateroconal fascia
LK	Left kidney
LgD	Longissimus dorsi muscle
LtD	Latissimus dorsi muscle
MeF	Mesenteric fat
Mf	Multifidus muscle
PD	Pancreatic duct
PF	Perirenal fat
PH	Pancreas, head

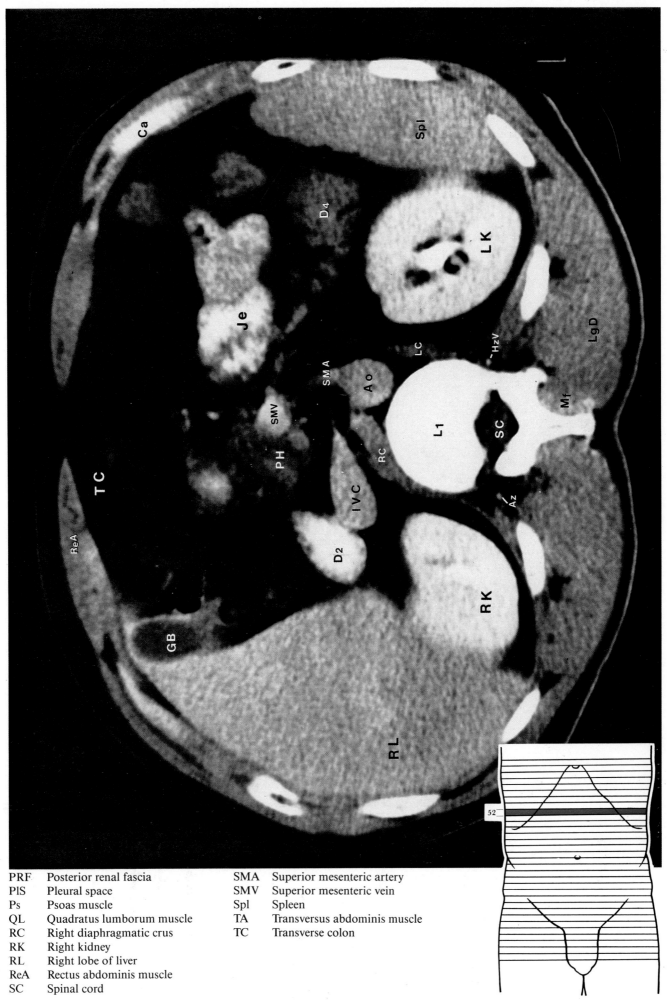

PRF	Posterior renal fascia
PlS	Pleural space
Ps	Psoas muscle
QL	Quadratus lumborum muscle
RC	Right diaphragmatic crus
RK	Right kidney
RL	Right lobe of liver
ReA	Rectus abdominis muscle
SC	Spinal cord

SMA	Superior mesenteric artery
SMV	Superior mesenteric vein
Spl	Spleen
TA	Transversus abdominis muscle
TC	Transverse colon

Plate 53. TRANSVERSE ABDOMEN

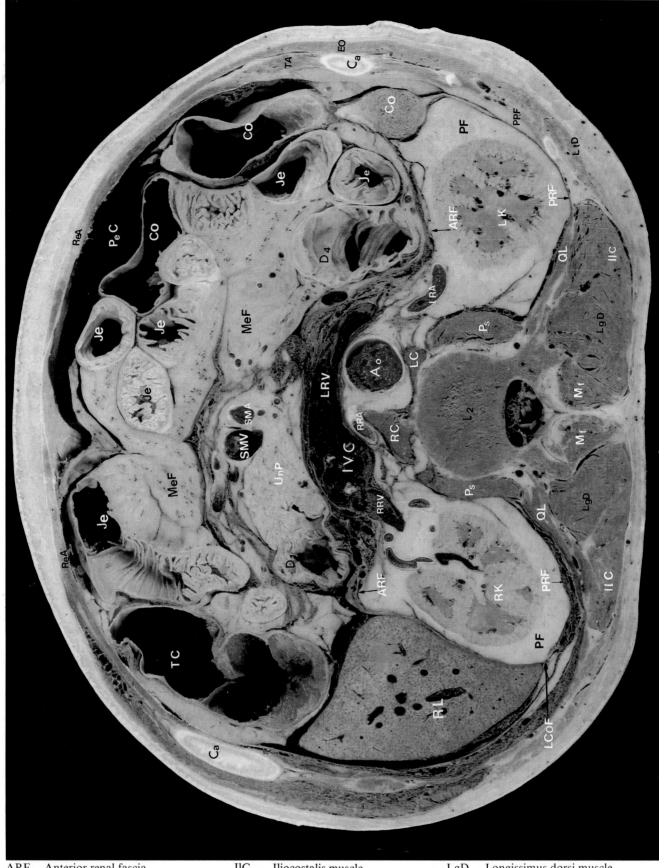

ARF	Anterior renal fascia	IlC	Iliocostalis muscle	LgD	Longissimus dorsi muscle
Ao	Aorta	Je	Jejunum	LtD	Latissimus dorsi muscle
Ca	Cartilage	L1-2	L1-2 intervertebral disc	MeF	Mesenteric fat
Co	Colon	L2	L2 vertebral body	Mf	Multifidus muscle
D1	First portion of duodenum	LC	Left diaphragmatic crus	PF	Perirenal fat
D2	Second portion of duodenum	LCoF	Lateroconal fascia	PPF	Posterior pararenal fat
D4	Fourth portion of duodenum	LK	Left kidney	PRF	Posterior renal fascia
EO	External oblique muscle	LRA	Left renal artery	PeC	Peritoneal cavity
IVC	Inferior vena cava	LRV	Left renal vein	Ps	Psoas muscle

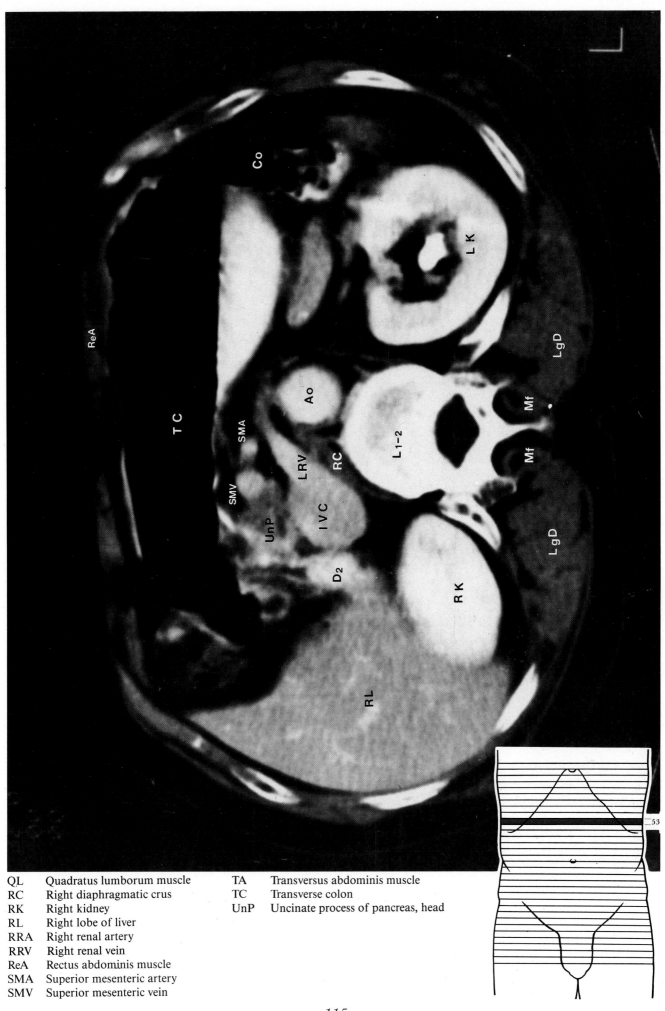

QL	Quadratus lumborum muscle		TA	Transversus abdominis muscle
RC	Right diaphragmatic crus		TC	Transverse colon
RK	Right kidney		UnP	Uncinate process of pancreas, head
RL	Right lobe of liver			
RRA	Right renal artery			
RRV	Right renal vein			
ReA	Rectus abdominis muscle			
SMA	Superior mesenteric artery			
SMV	Superior mesenteric vein			

Plate 54. TRANSVERSE ABDOMEN

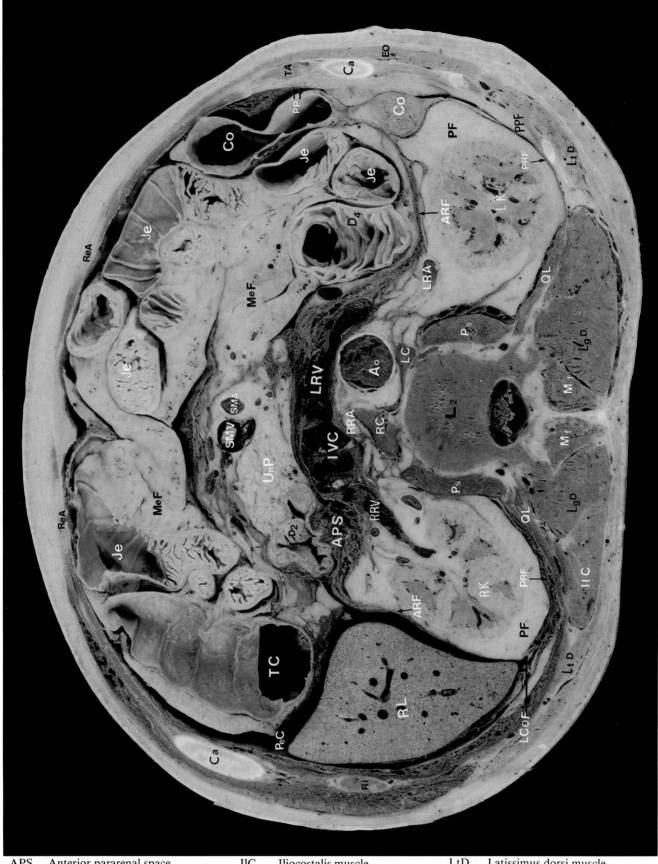

APS	Anterior pararenal space	IIC	Iliocostalis muscle	LtD	Latissimus dorsi muscle
ARF	Anterior renal fascia	Je	Jejunum	MeF	Mesenteric fat
Ao	Aorta	L2	L2 vertebral body	Mf	Multifidus muscle
Ca	Cartilage	LC	Left diaphragmatic crus	PF	Perirenal fat
Co	Colon	LCoF	Lateroconal fascia	PP	Parietal peritoneum
D2	Second portion of duodenum	LK	Left kidney	PPF	Posterior pararenal fat
D4	Fourth portion of duodenum	LRA	Left renal artery	PRF	Posterior renal fascia
EO	External oblique muscle	LRV	Left renal vein	PeC	Peritoneal cavity
IVC	Inferior vena cava	LgD	Longissimus dorsi muscle	Ps	Psoas muscle

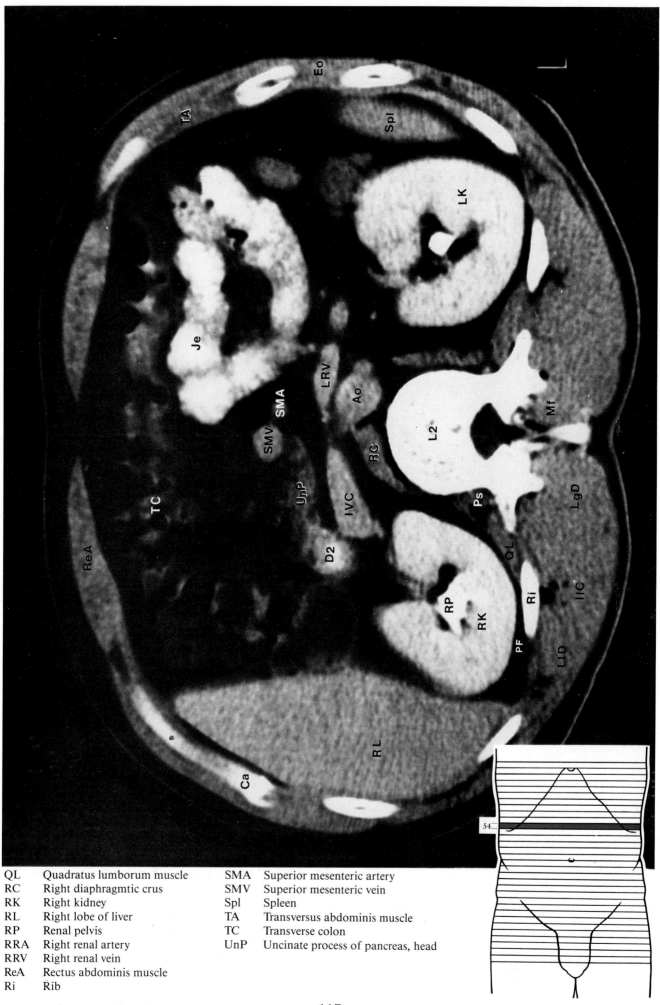

QL	Quadratus lumborum muscle	SMA	Superior mesenteric artery
RC	Right diaphragmtic crus	SMV	Superior mesenteric vein
RK	Right kidney	Spl	Spleen
RL	Right lobe of liver	TA	Transversus abdominis muscle
RP	Renal pelvis	TC	Transverse colon
RRA	Right renal artery	UnP	Uncinate process of pancreas, head
RRV	Right renal vein		
ReA	Rectus abdominis muscle		
Ri	Rib		

Plate 55. TRANSVERSE ABDOMEN

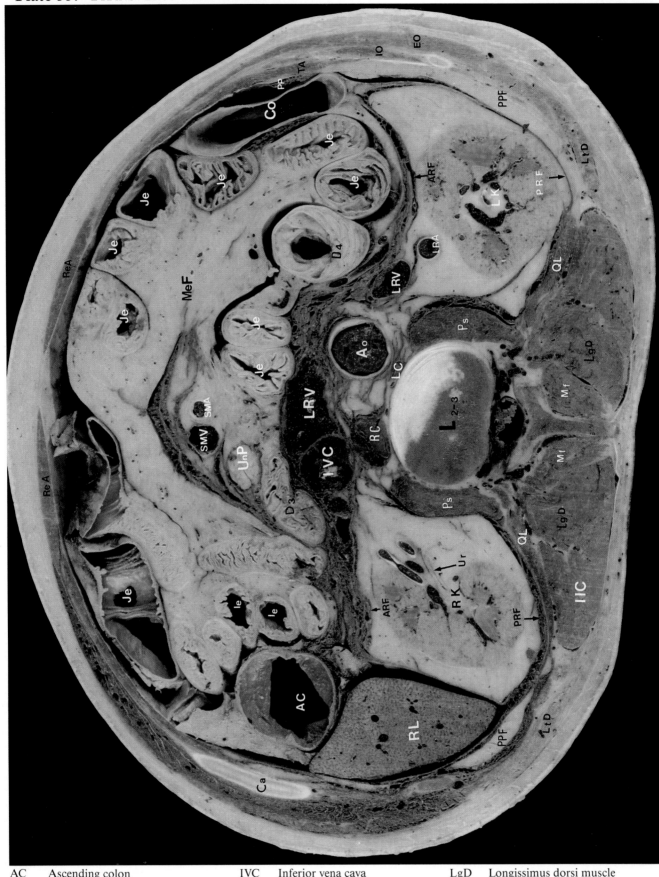

AC	Ascending colon	IVC	Inferior vena cava	LgD	Longissimus dorsi muscle
ARF	Anterior renal fascia	Ie	Ileum	LtD	Latissimus dorsi muscle
Ao	Aorta	IlC	Iliocostalis muscle	MeF	Mesenteric fat
Ca	Cartilage	Je	Jejunum	Mf	Multifidus muscle
Co	Colon	L2-3	L2-3 intervertebral disc	PP	Parietal peritoneum
D3	Third portion of duodenum	LC	Left diaphragmatic crus	PPF	Posterior pararenal fat
D4	Fourth portion of duodenum	LK	Left kidney	PRF	Posterior renal fascia
EO	External oblique muscle	LRA	Left renal artery	Ps	Psoas muscle
IO	Internal oblique muscle	LRV	Left renal vein	QL	Quadratus lumborum muscle

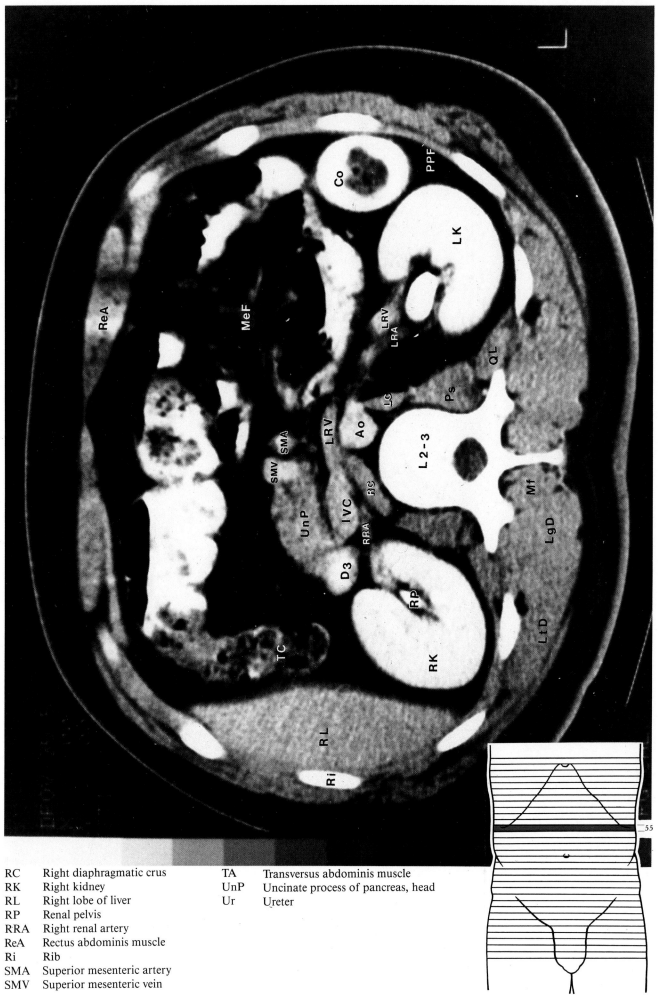

RC	Right diaphragmatic crus
RK	Right kidney
RL	Right lobe of liver
RP	Renal pelvis
RRA	Right renal artery
ReA	Rectus abdominis muscle
Ri	Rib
SMA	Superior mesenteric artery
SMV	Superior mesenteric vein

TA	Transversus abdominis muscle
UnP	Uncinate process of pancreas, head
Ur	Ureter

Plate 56. TRANSVERSE ABDOMEN

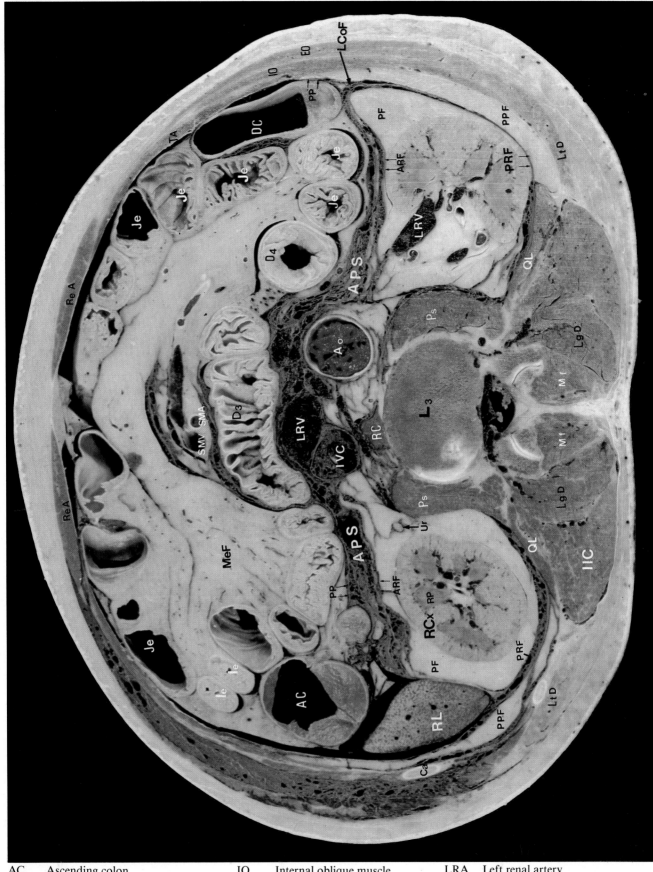

AC	Ascending colon	IO	Internal oblique muscle	LRA	Left renal artery
APS	Anterior pararenal space	IVC	Inferior vena cava	LRV	Left renal vein
ARF	Anterior renal fascia	Ie	Ileum	LgD	Longissimus dorsi muscle
Ao	Aorta	IlC	Iliocostalis muscle	LtD	Latissimus dorsi muscle
Ca	Cartilage	Je	Jejunum	MeF	Mesenteric fat
D3	Third portion of duodenum	L3	L3 vertebral body	Mf	Multifidus muscle
D4	Fourth portion of duodenum	LC	Left diaphragmatic crus	PF	Perirenal fat
DC	Descending colon	LCoF	Lateroconal fascia	PP	Parietal peritoneum
EO	External oblique muscle	LK	Left kidney	PPF	Posterior pararenal fat

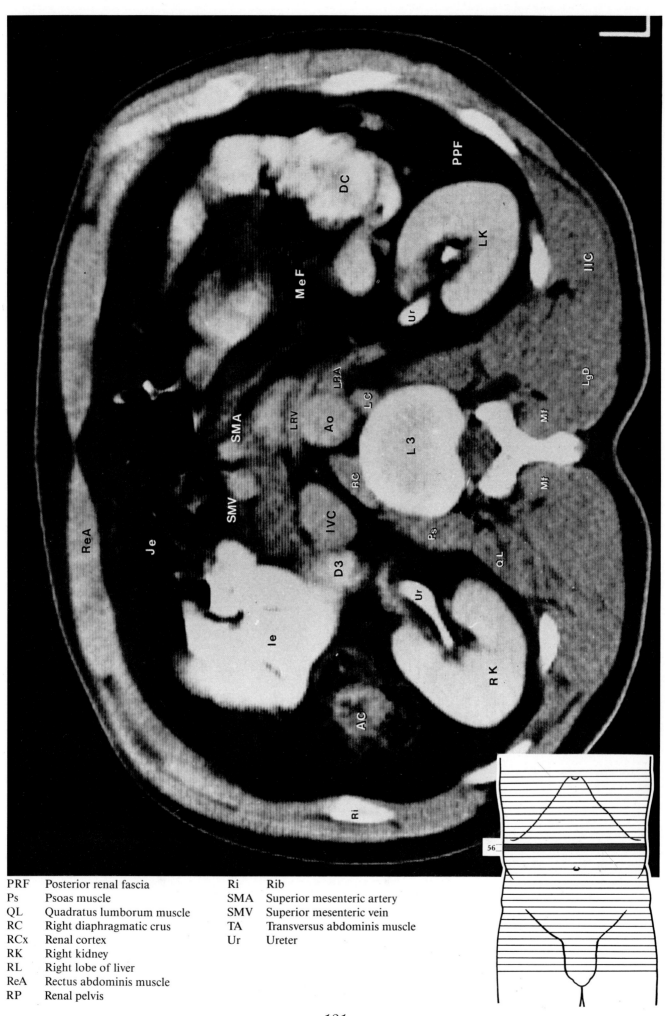

PRF	Posterior renal fascia	Ri	Rib
Ps	Psoas muscle	SMA	Superior mesenteric artery
QL	Quadratus lumborum muscle	SMV	Superior mesenteric vein
RC	Right diaphragmatic crus	TA	Transversus abdominis muscle
RCx	Renal cortex	Ur	Ureter
RK	Right kidney		
RL	Right lobe of liver		
ReA	Rectus abdominis muscle		
RP	Renal pelvis		

Plate 57. TRANSVERSE ABDOMEN

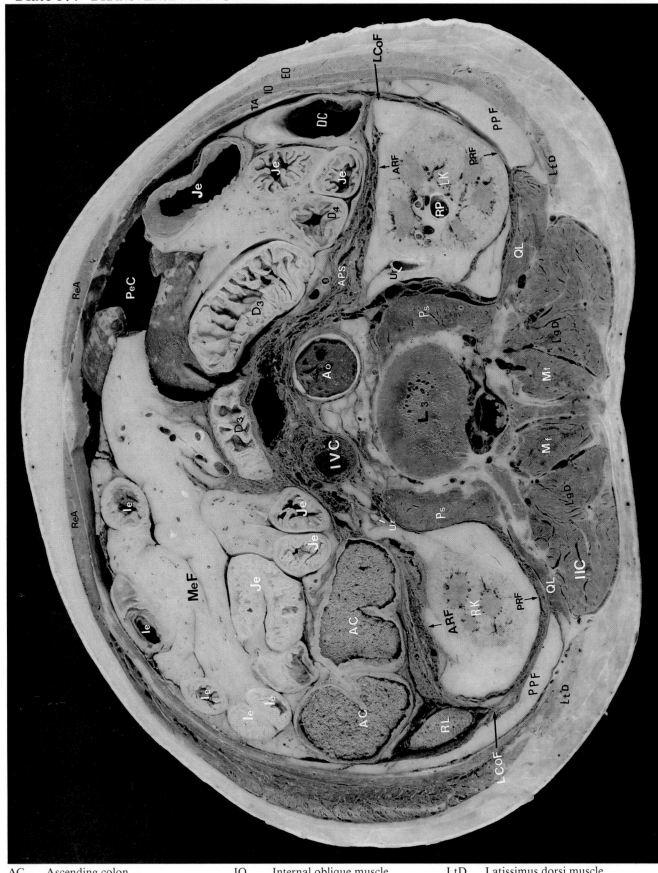

AC	Ascending colon	IO	Internal oblique muscle	LtD	Latissimus dorsi muscle
APS	Anterior pararenal space	IVC	Inferior vena cava	MeF	Mesenteric fat
ARF	Anterior renal fascia	Ie	Ileum	Mf	Multifidus muscle
Ao	Aorta	IlC	Iliocostalis muscle	PPF	Posterior pararenal fat
Ca	Cartilage	Je	Jejunum	PRF	Posterior renal fascia
D3	Third portion of duodenum	L3	L3 vertebral body	PeC	Peritoneal cavity
D4	Fourth portion of duodenum	LCoF	Lateroconal fascia	Ps	Psoas muscle
DC	Descending colon	LK	Left kidney	QL	Quadratus lumborum muscle
EO	External oblique muscle	LgD	Longissimus dorsi muscle	RK	Right kidney

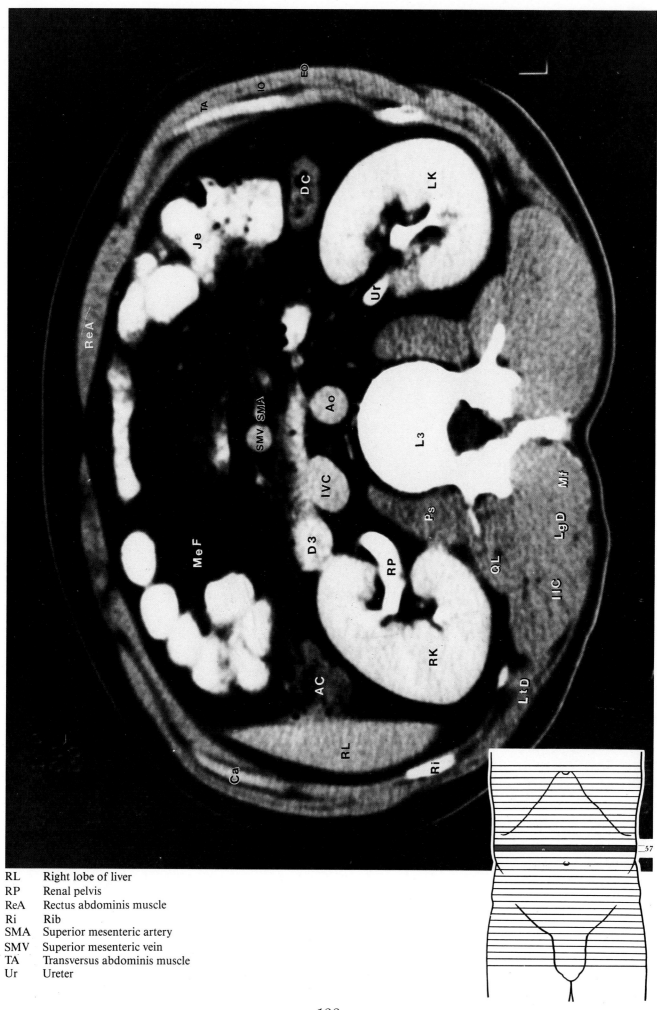

RL   Right lobe of liver
RP   Renal pelvis
ReA Rectus abdominis muscle
Ri   Rib
SMA Superior mesenteric artery
SMV Superior mesenteric vein
TA   Transversus abdominis muscle
Ur   Ureter

Plate 58. TRANSVERSE ABDOMEN

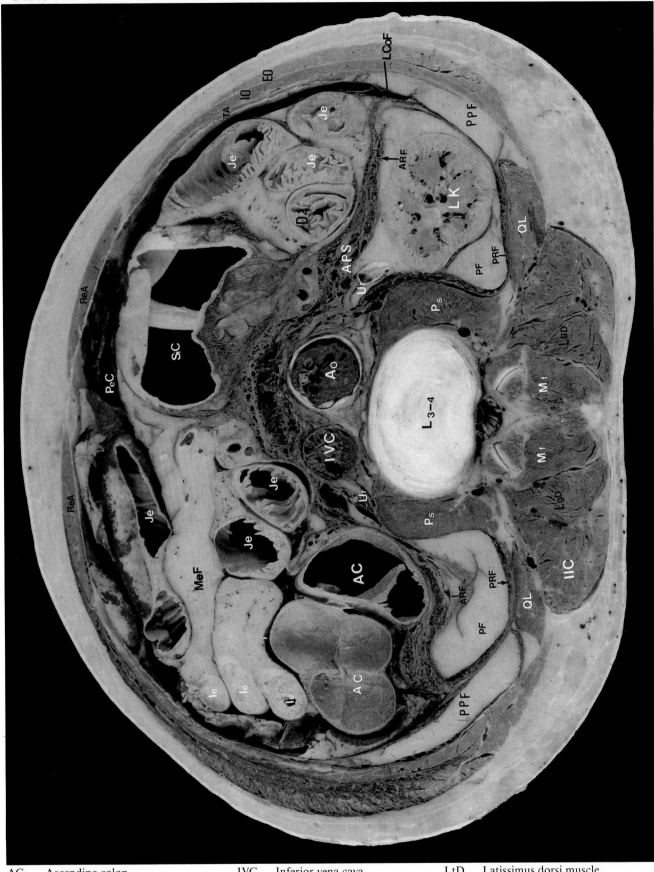

AC	Ascending colon	IVC	Inferior vena cava	LtD	Latissimus dorsi muscle
APS	Anterior pararenal space	Ie	Ileum	MeF	Mesenteric fat
ARF	Anterior renal fascia	IlC	Iliocostalis muscle	Mf	Multifidus muscle
Ao	Aorta	Je	Jejunum	PF	Perirenal fat
D3	Third portion of duodenum	L3-4	L3-4 intervertebral disc	PPF	Posterior pararenal fat
D4	Fourth portion of duodenum	LC	Left diaphragmatic crus	PRF	Posterior renal fascia
DC	Descending colon	LCoF	Lateroconal fascia	PeC	Peritoneal cavity
EO	External oblique muscle	LK	Left kidney	Ps	Psoas muscle
IO	Internal oblique muscle	LgD	Longissimus dorsi muscle	QL	Quadratus lumborum muscle

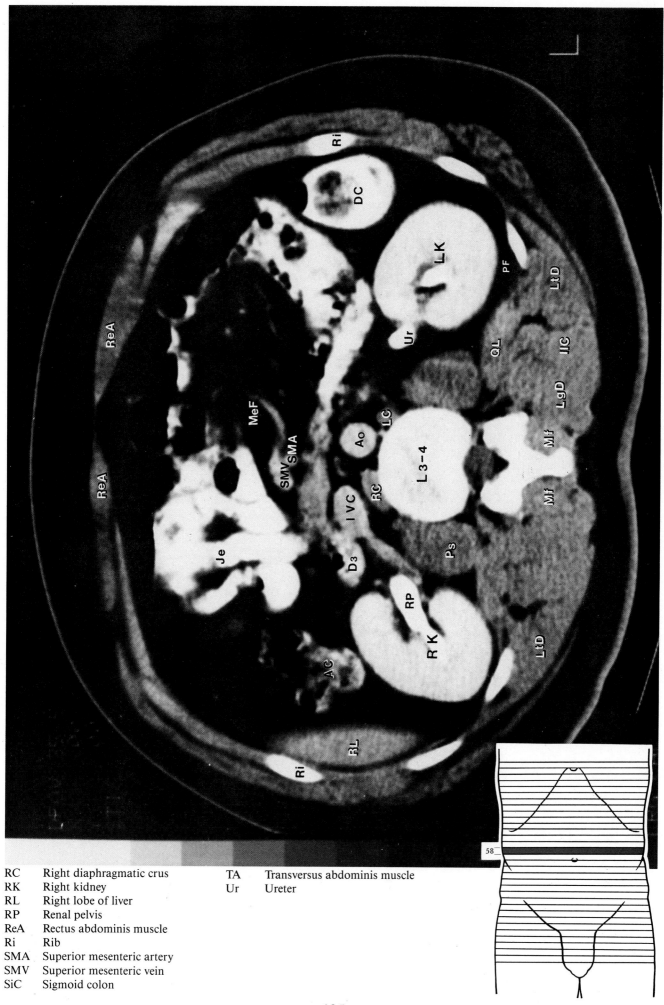

Plate 59. TRANSVERSE ABDOMEN

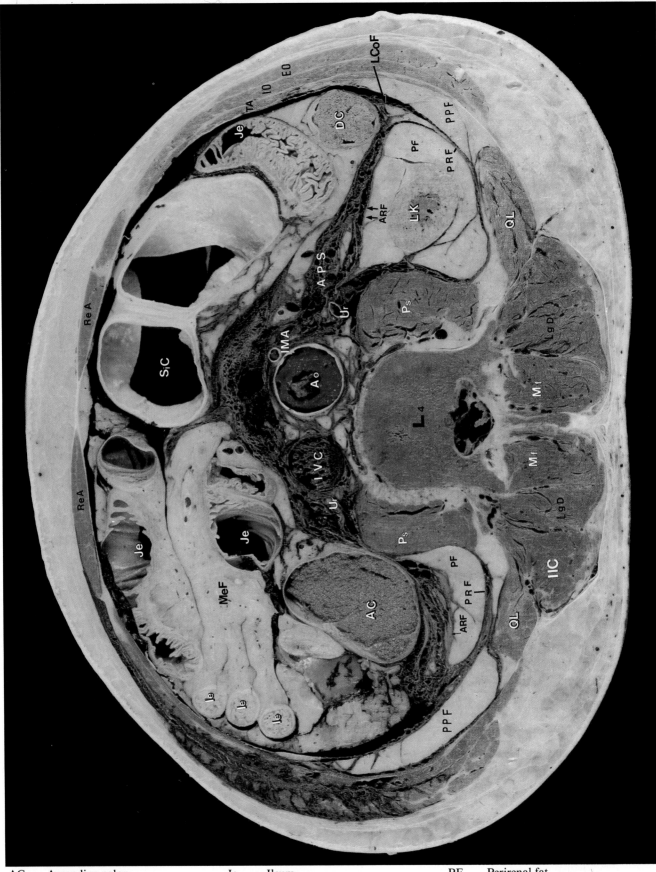

AC	Ascending colon	Ie	Ileum	PF	Perirenal fat
APS	Anterior pararenal space	IIC	Iliocostalis muscle	PPF	Posterior pararenal fat
ARF	Anterior renal fascia	Je	Jejunum	PRF	Posterior renal fascia
Ao	Aorta	L4	L4 vertebral body	Ps	Psoas muscle
DC	Descending colon	LCoF	Lateroconal fascia	QL	Quadratus lumborum muscle
EO	External oblique muscle	LK	Left kidney	RK	Right kidney
IMA	Inferior mesenteric artery	LgD	Longissimus dorsi muscle	RP	Renal pelvis
IO	Internal oblique muscle	MeF	Mesenteric fat	ReA	Rectus abdominis muscle
IVC	Inferior vena cava	Mf	Multifidus muscle	SMA	Superior mesenteric artery

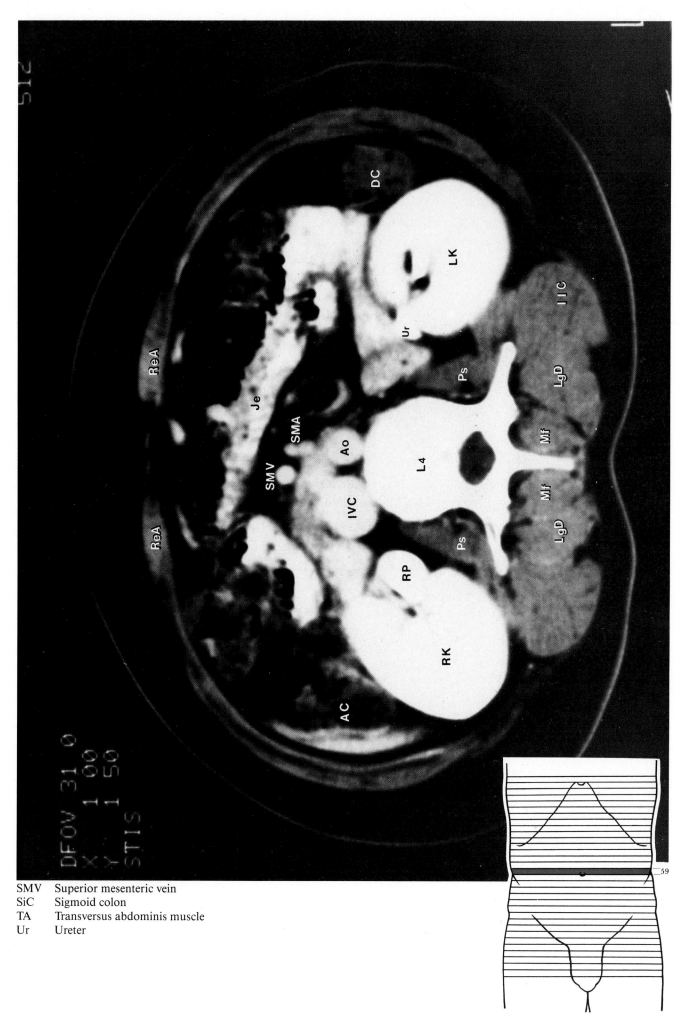

SMV   Superior mesenteric vein
SiC   Sigmoid colon
TA   Transversus abdominis muscle
Ur   Ureter

Plate 60. TRANSVERSE PELVIS-MALE

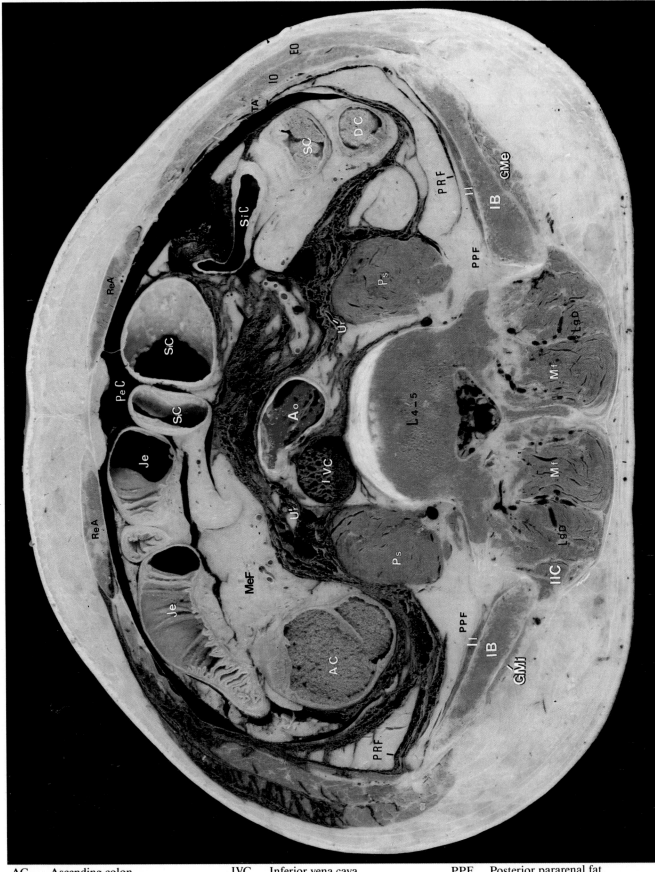

AC	Ascending colon	IVC	Inferior vena cava	PPF	Posterior pararenal fat
Ao	Aorta	IlC	Iliocostalis muscle	PRF	Posterior renal fascia
DC	Descending colon	Il	Iliacus muscle	PeC	Peritoneal cavity
EO	External oblique muscle	Je	Jejunum	Ps	Psoas muscle
GMe	Gluteus medius muscle	L4	L4 vertebral body	QL	Quadratus lumborum muscle
GMi	Gluteus minimus muscle	L4-5	L4-5 intervertebral disc	ReA	Rectus abdominis muscle
IB	Iliac bone	LgD	Longissimus dorsi muscle	SiC	Sigmoid colon
IMA	Inferior mesenteric artery	MeF	Mesenteric fat	TA	Transversus abdominis muscle
IO	Internal oblique muscle	Mf	Multifidus muscle	Ur	Ureter

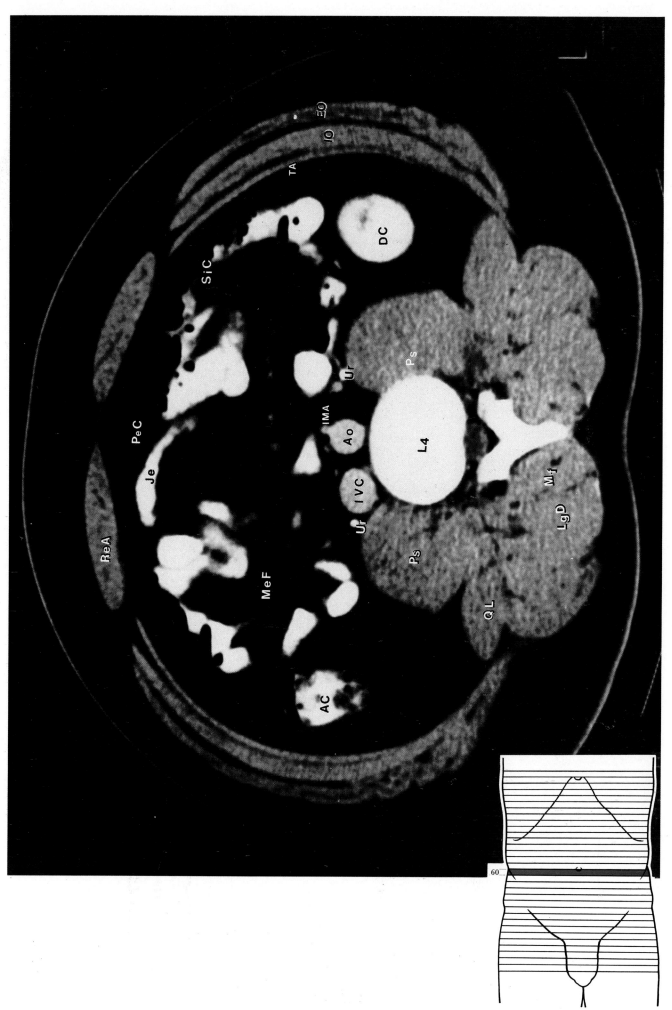

Plate 61. TRANSVERSE PELVIS-MALE

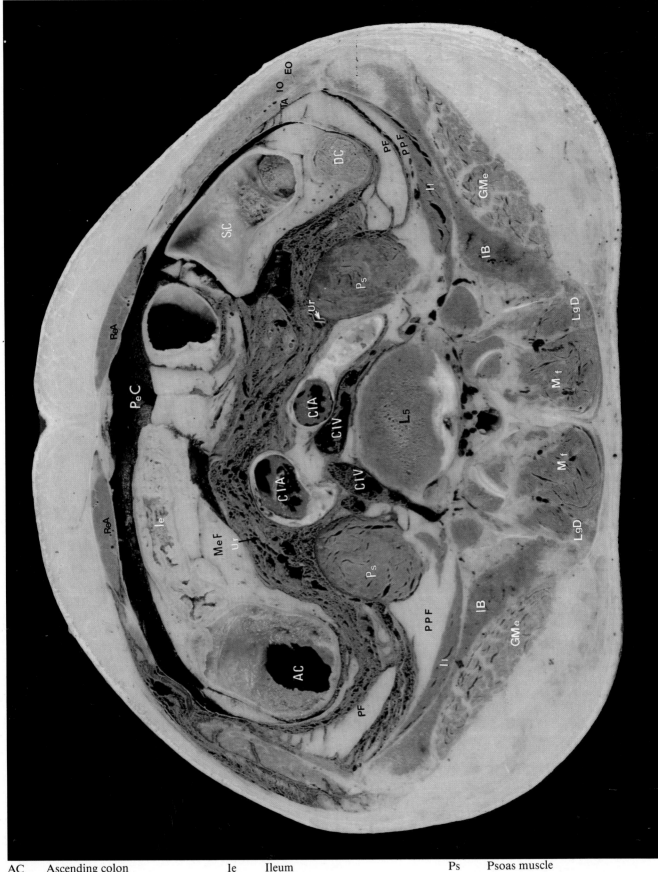

AC	Ascending colon	Ie	Ileum	Ps	Psoas muscle
CIA	Common iliac artery	Il	Iliacus muscle	ReA	Rectus abdominis muscle
CIV	Common iliac vein	L5	L5 vertebral body	SiC	Sigmoid colon
DC	Descending colon	LgD	Longissimus dorsi muscle	TA	Transversus abdominis muscle
EO	External oblique muscle	MeF	Mesenteric fat	Ur	Ureter
GMe	Gluteus medius muscle	Mf	Multifidus muscle		
IB	Iliac bone	PF	Perirenal fat		
IO	Internal oblique muscle	PPF	Posterior pararenal fat		
IVC	Inferior vena cava	PeC	Peritoneal cavity		

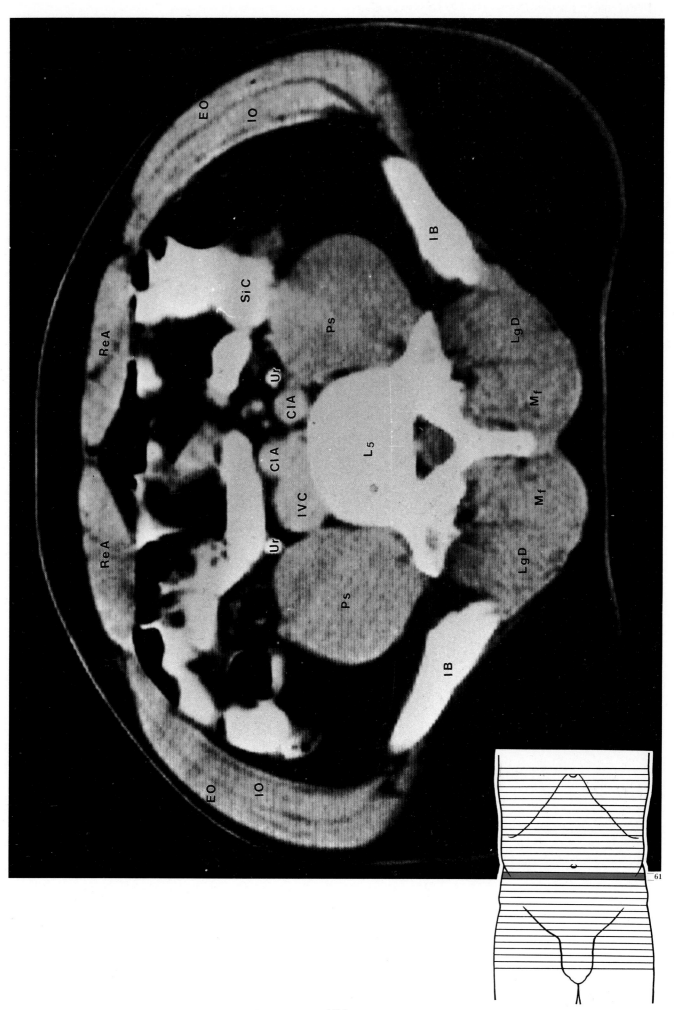

Plate 62. TRANSVERSE PELVIS-MALE

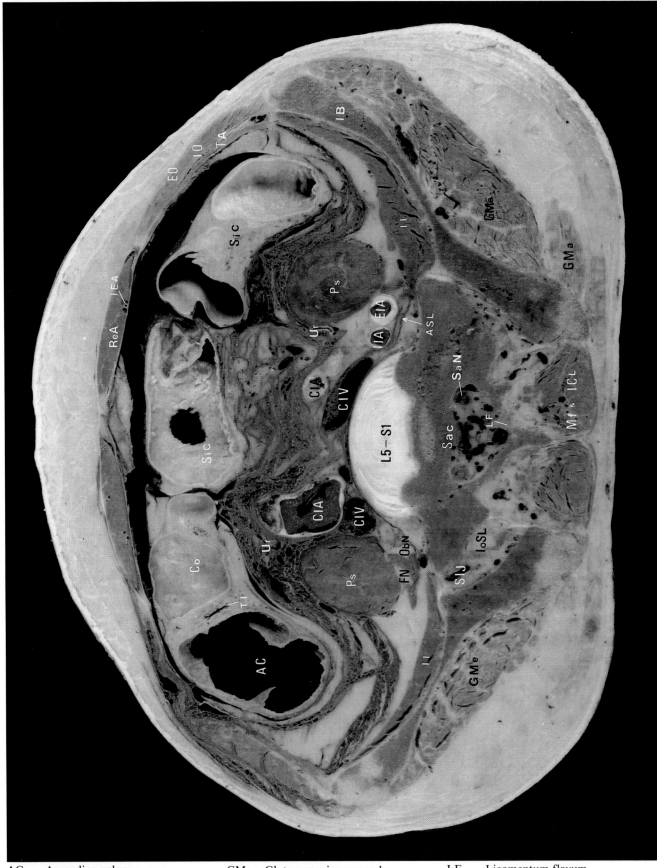

AC	Ascending colon	GMa	Gluteus maximus muscle	LF	Ligamentum flavum
ASL	Anterior sacroiliac ligament	GMe	Gluteus medius muscle	Mf	Multifidus muscle
CIA	Common iliac artery	IB	Iliac bone	ObN	Obturator nerve
CIV	Common iliac vein	ICL	Iliocostalis lumborum muscle	Ps	Psoas muscle
Co	Colon	IIA	Internal iliac artery	ReA	Rectus abdominis muscle
DC	Descending colon	IO	Internal oblique muscle	SIJ	Sacroiliac joint
EIA	External iliac artery	Il	Iliacus muscle	SaN	Sacral nerve
EO	External oblique muscle	IoSL	Interosseous sacroiliac ligament	Sac	Sacrum
FN	Femoral nerve	L5-S1	L5-S1 intervertebral disc	SiC	Sigmoid colon

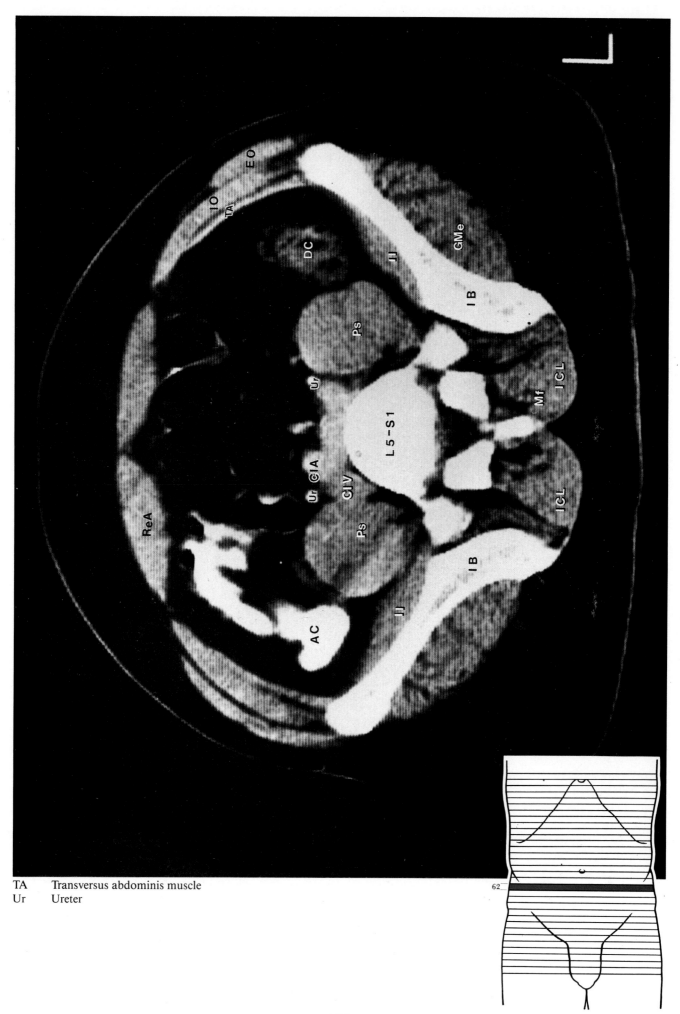

TA   Transversus abdominis muscle
Ur   Ureter

Plate 63. TRANSVERSE PELVIS-MALE

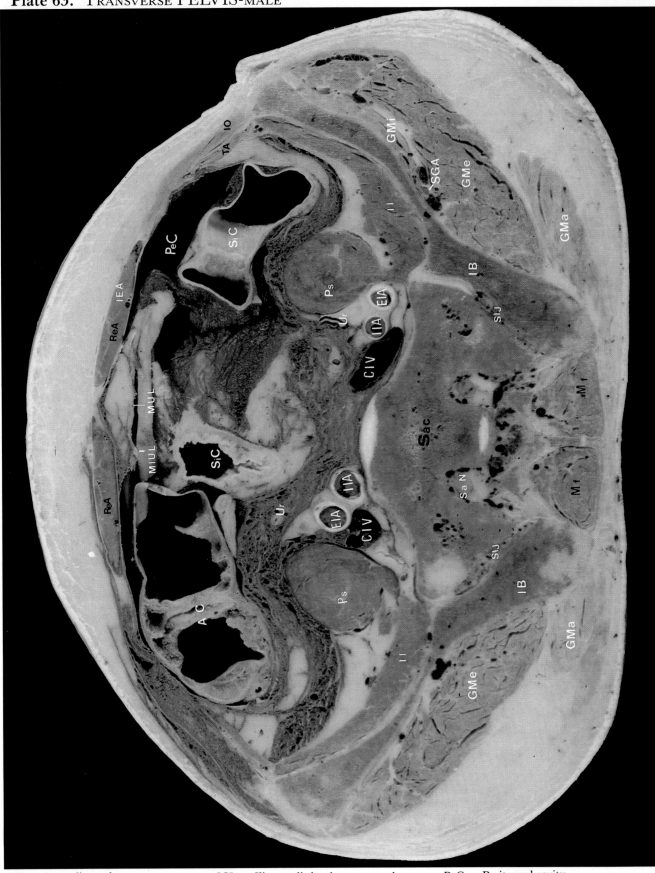

AC	Ascending colon	ICL	Iliocostalis lumborum muscle	PeC	Peritoneal cavity
CIA	Common iliac artery	IEA	Inferior epigastric artery	Ps	Psoas muscle
CIV	Common iliac vein	IIA	Internal iliac artery	ReA	Rectus abdominis muscle
EIA	External iliac artery	IO	Internal oblique muscle	SGA	Superior gluteal artery
EO	External oblique muscle	ll	Iliacus muscle	SIJ	Sacroiliac joint
GMa	Gluteus maximus muscle	L5-S1	L5-S1 intervertebral disc	SaN	Sacral nerve
GMe	Gluteus medius muscle	MUL	Median umbilical ligament	Sac	Sacrum
GMi	Gluteus minimus muscle	Mf	Multifidus mescle	SiC	Sigmoid colon
IB	Iliac bone	MlUL	Medial umbilical ligament	TA	Transversus abdominis muscle

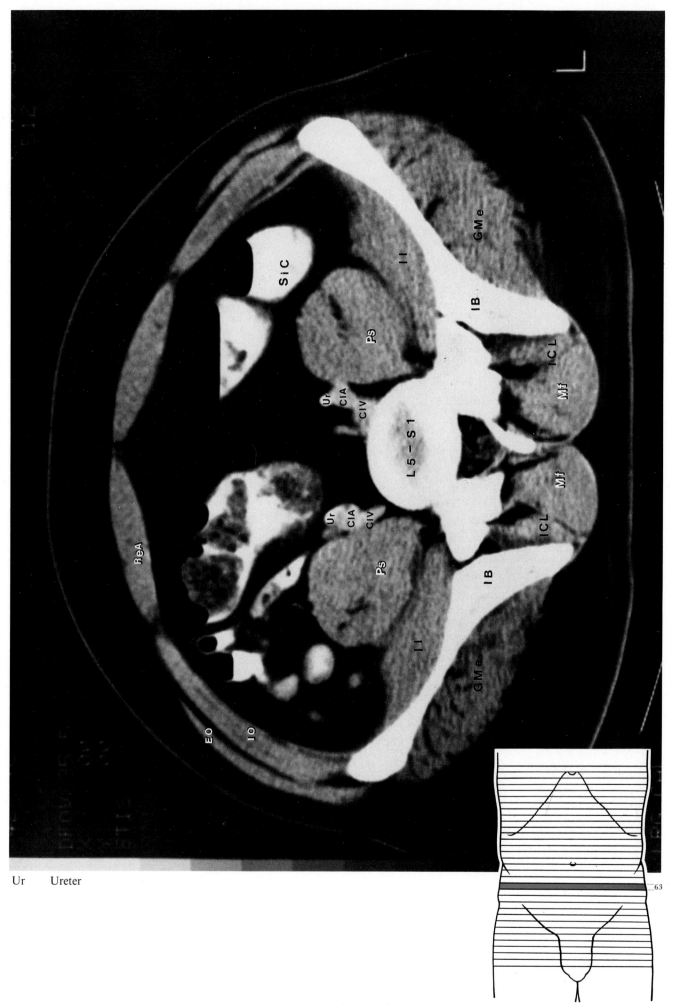

Ur   Ureter

Plate 64. TRANSVERSE PELVIS-MALE

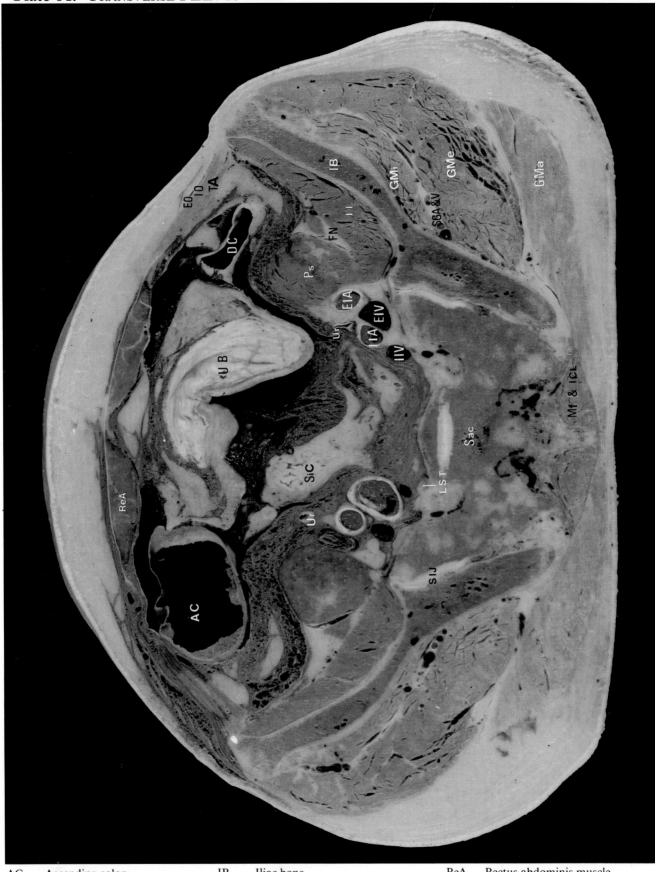

AC	Ascending colon	IB	Iliac bone	ReA	Rectus abdominis muscle
DC	Descending colon	ICL	Iliocostalis lumborum muscle	SGA	Superior gluteal artery
EIA	External iliac artery	IIA	Internal iliac artery	SGV	Superior gluteal vein
EIV	External iliac vein	IIV	Internal iliac vein	SIJ	Sacroiliac joint
EO	External oblique muscle	IO	Internal oblique muscle	Sac	Sacrum
FN	Femoral nerve	Il	Iliacus muscle	SiC	Sigmoid colon
GMa	Gluteus maximus muscle	LST	Lumbosacral trunk	TA	Transversus abdominis muscle
GMe	Gluteus medius muscle	Mf	Multifidus muscle	UB	Urinary bladder
GMi	Gluteus minimus muscle	Ps	Psoas muscle	Ur	Ureter

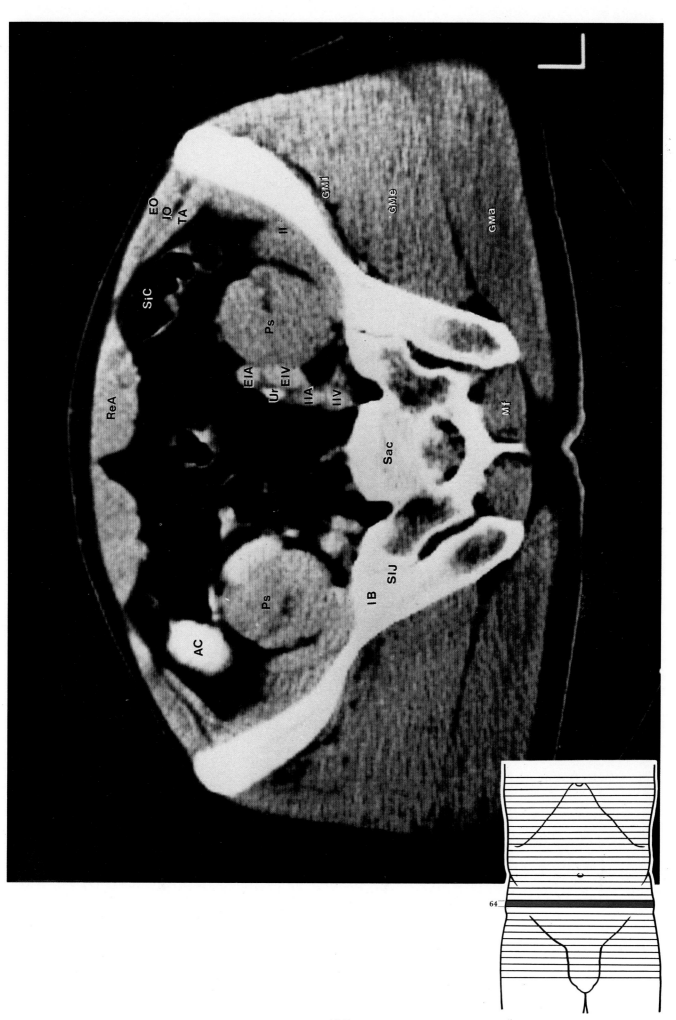

Plate 65. Transverse PELVIS-male

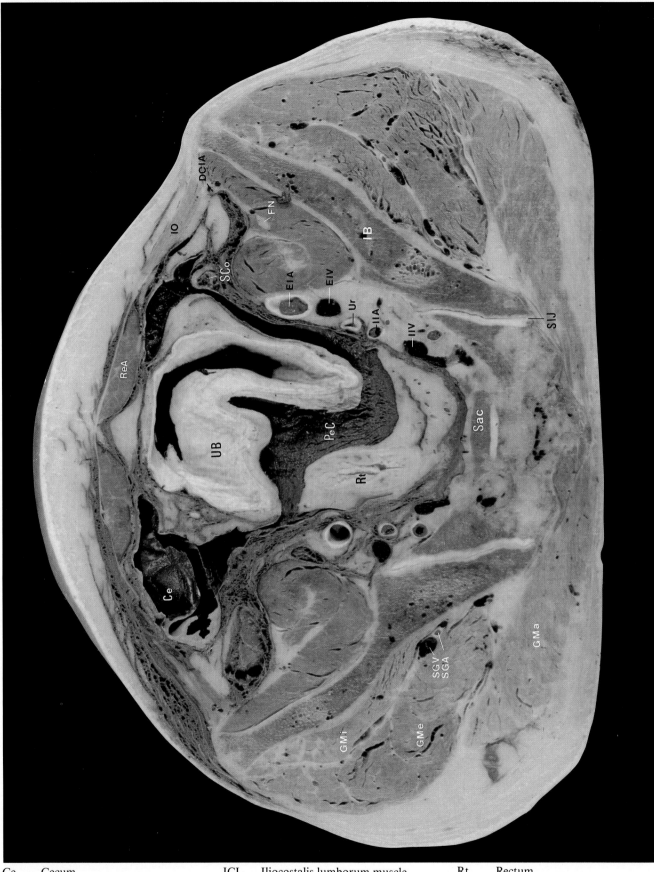

Ce	Cecum	ICL	Iliocostalis lumborum muscle	Rt	Rectum
DCIA	Deep circumflex iliac artery	IIA	Internal iliac artery	SCo	Spermatic cord
EIA	External iliac artery	IIV	Internal iliac vein	SGA	Superior gluteal artery
EIV	External iliac vein	IO	Internal oblique muscle	SGV	Superior gluteal vein
FN	Femoral nerve	Il	Iliacus muscle	SIJ	Sacroiliac joint
GMa	Gluteus maximus muscle	Mf	Multifidus muscle	Sac	Sacrum
GMe	Gluteus medius muscle	PeC	Peritoneal cavity	UB	Urinary bladder
GMi	Gluteus minimus muscle	Ps	Psoas muscle	Ur	Ureter
IB	Iliac bone	ReA	Rectus abdominis muscle		

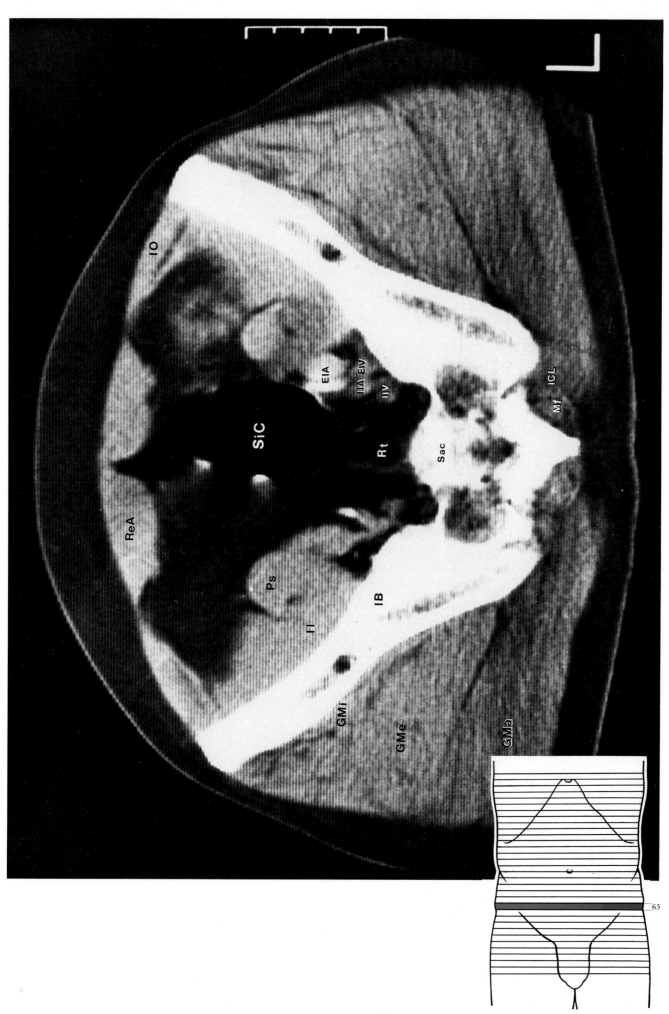

Plate 66. TRANSVERSE PELVIS-MALE

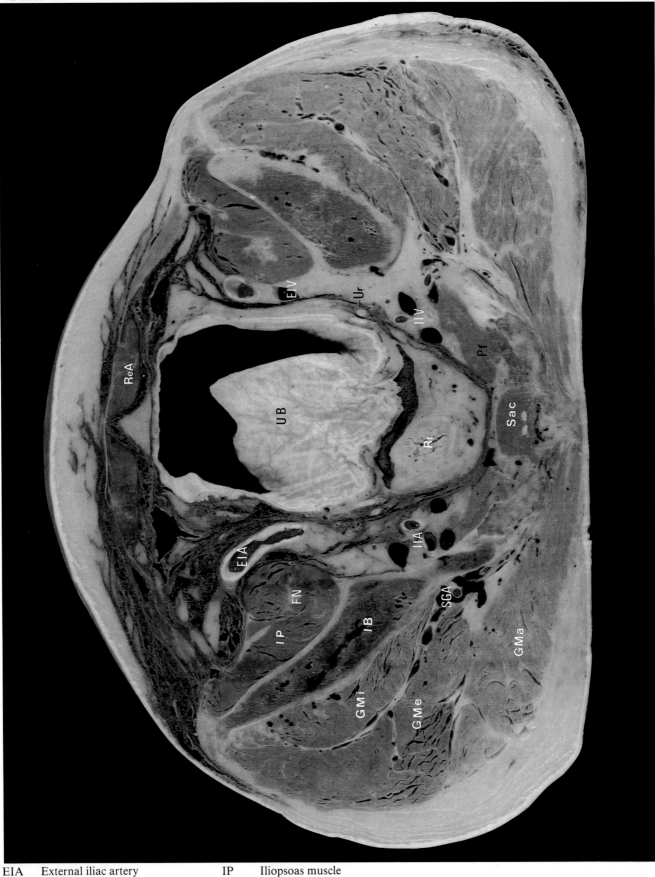

EIA	External iliac artery	IP	Iliopsoas muscle
EIV	External iliac vein	Pf	Piriformis muscle
FN	Femoral nerve	ReA	Rectus abdominis muscle
GMa	Gluteus maximus muscle	Rt	Rectum
GMe	Gluteus medius muscle	SGA	Superior gluteal artery
GMi	Gluteus minimus muscle	Sac	Sacrum
IB	Iliac bone	Sar	Sartorius muscle
IIA	Internal iliac artery	UB	Urinary bladder
IIV	Internal iliac vein	Ur	Ureter

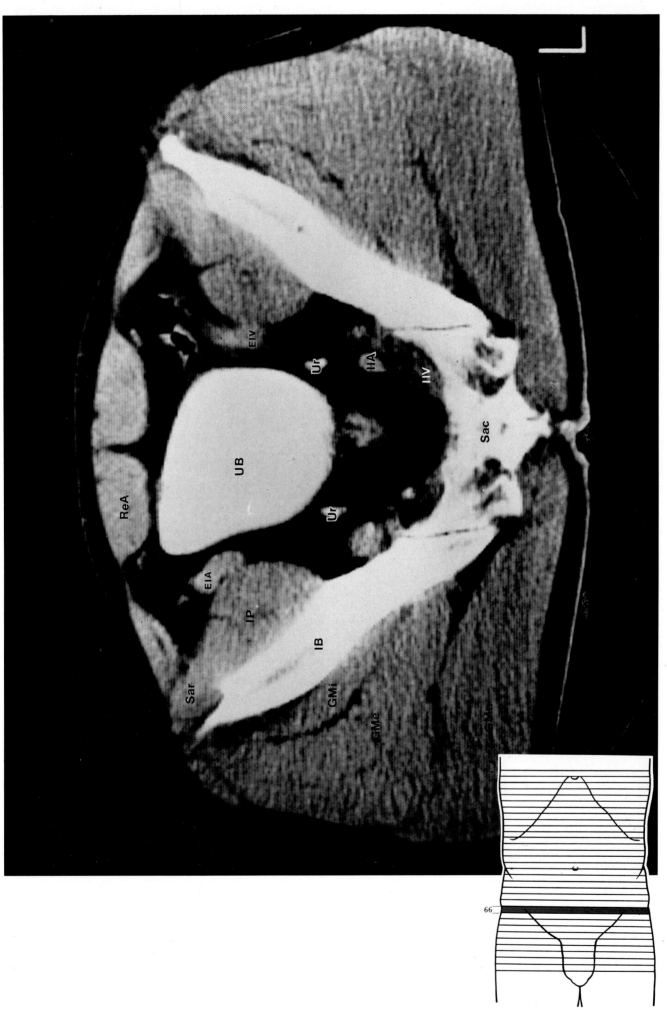

Plate 67. TRANSVERSE PELVIS-MALE

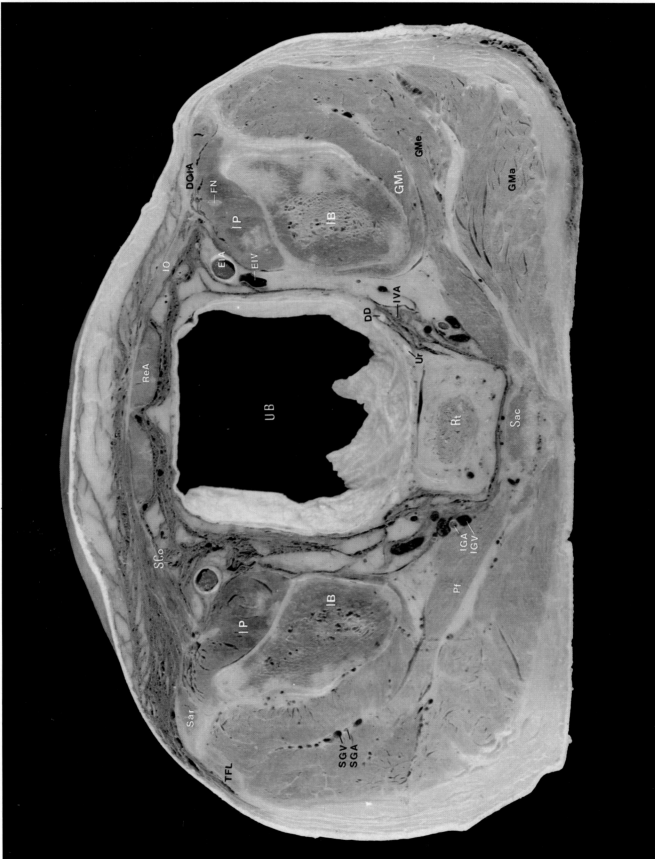

DCIA	Deep circumflex iliac artery	IGA	Inferior gluteal artery	SGA	Superior gluteal artery
DD	Ductus deferens	IGV	Inferior gluteal vein	SGV	Superior gluteal vein
EIA	External iliac artery	IO	Internal oblique muscle	Sac	Sacrum
EIV	External iliac vein	IP	Iliopsoas muscle	Sar	Sartorius muscle
FN	Femoral nerve	IVA	Inferior vesical artery	TFL	Tensor fascia lata
GMa	Gluteus maximus muscle	Pf	Piriformis muscle	UB	Urinary bladder
GMe	Gluteus medius muscle	ReA	Rectus abdominis muscle	Ur	Ureter
GMi	Gluteus minimus muscle	Rt	Rectum		
IB	Iliac bone	SCo	Spermatic cord		

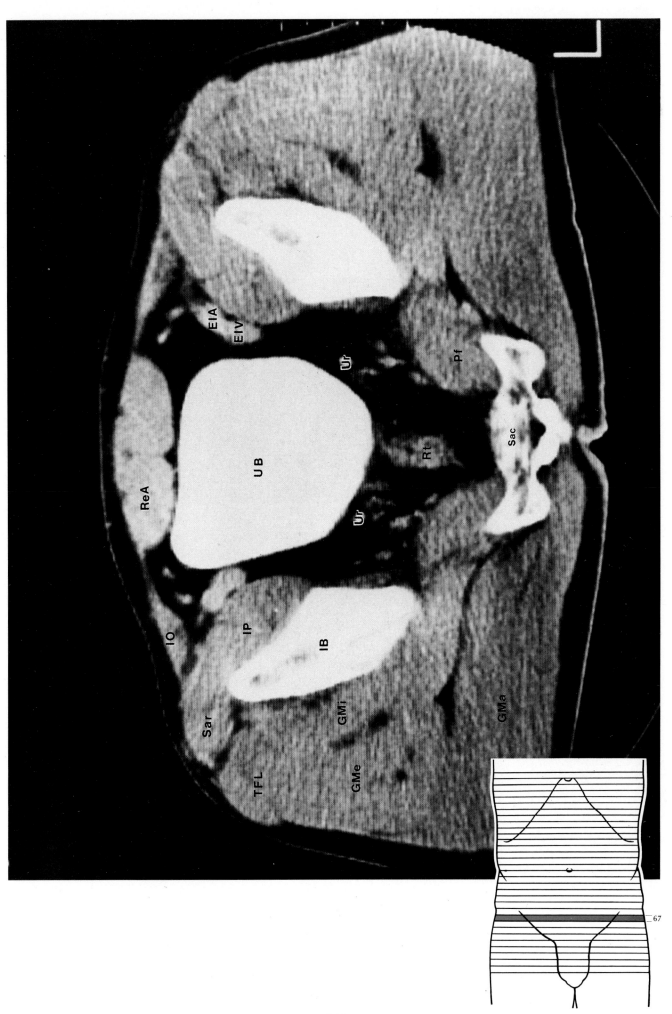

Plate 68. TRANSVERSE PELVIS-MALE

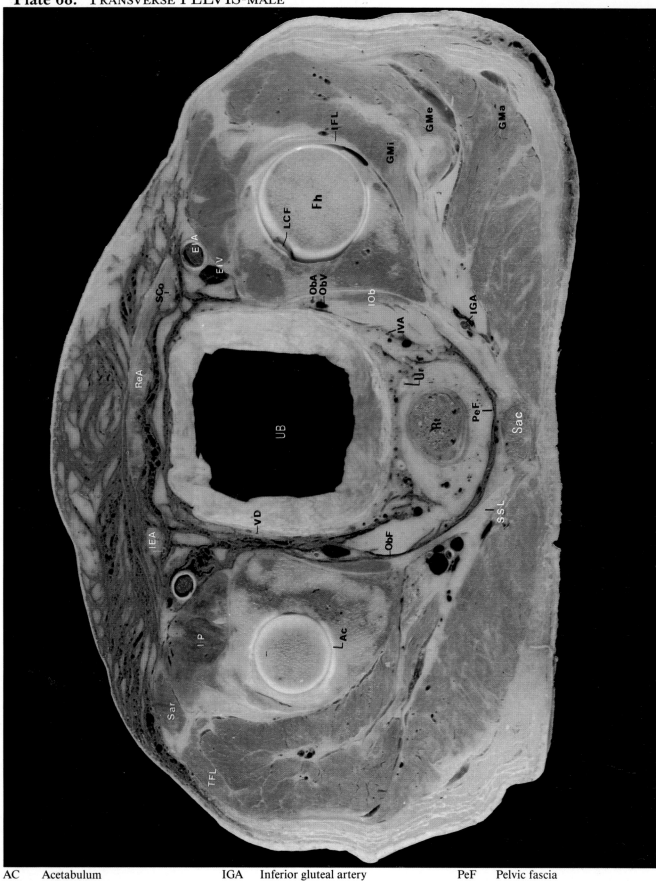

AC	Acetabulum	IGA	Inferior gluteal artery	PeF	Pelvic fascia
EIA	External iliac artery	IOb	Internal obturator muscle	ReA	Rectus abdominis muscle
EIV	External iliac vein	IP	Iliopsoas muscle	Rt	Rectum
Fh	Femur, head	IVA	Inferior vesical artery	SCo	Spermatic cord
GMa	Gluteus maximus muscle	LCF	Ligamentum capitis femoris	SSL	Sacrospinous ligament
GMe	Gluteus medius muscle	Ob	Obtuator muscle	Sac	Sacrum
GMi	Gluteus minimus muscle	ObA	Obturator artery	Sar	Sartorius muscle
IEA	Inferior epigastric artery	ObF	Obturator fascia	TFL	Tensor fascia lata
IFL	Iliofemoral ligament	ObV	Obturator vein	UB	Urinary bladder

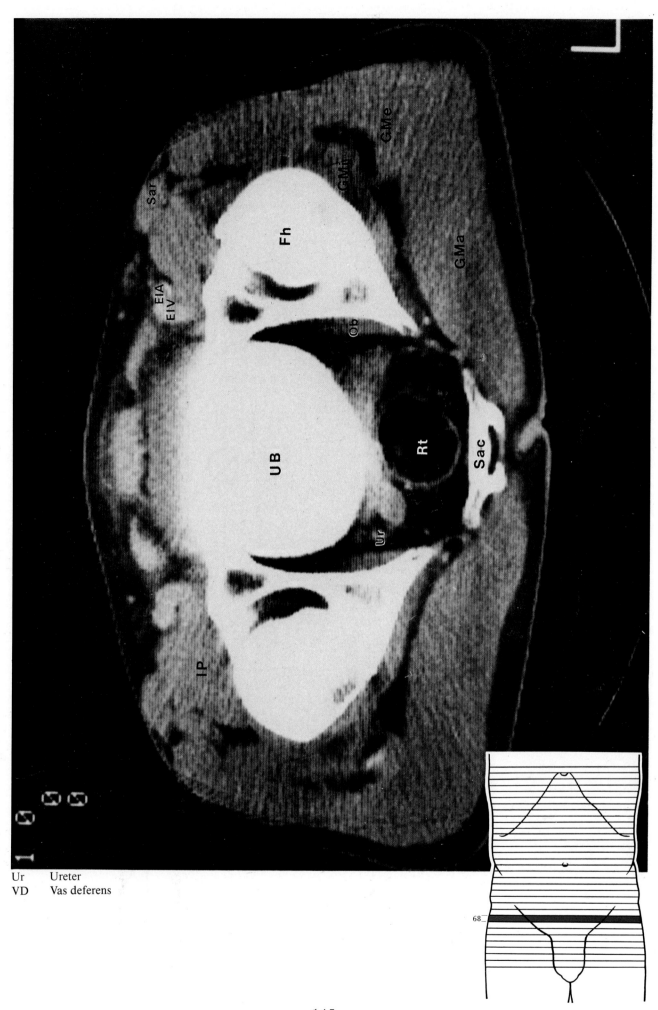

Ur   Ureter
VD   Vas deferens

Plate 69. TRANSVERSE PELVIS-MALE

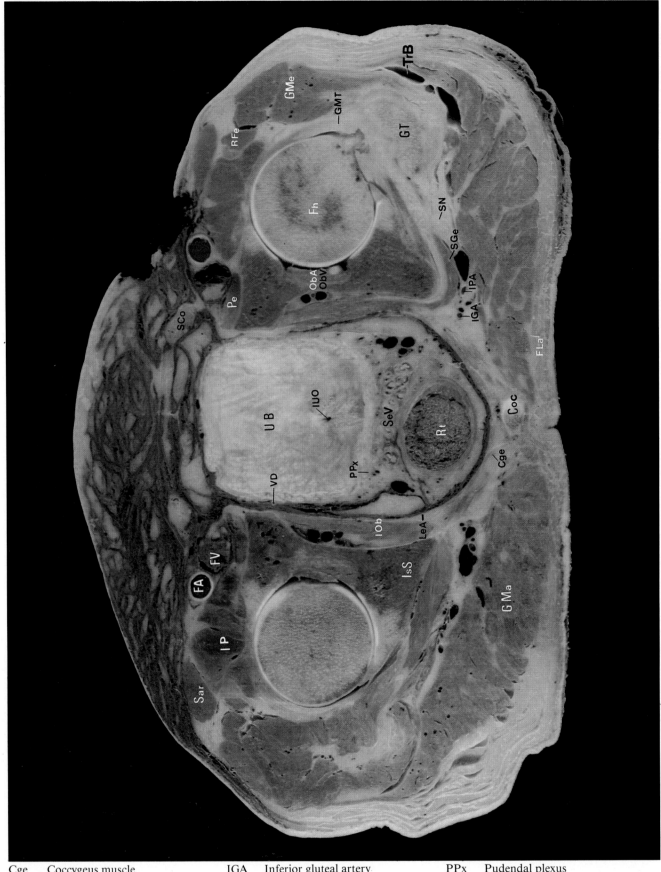

Cge	Coccygeus muscle	IGA	Inferior gluteal artery	PPx	Pudendal plexus
Coc	Coccyx	IOb	Internal obturator muscle	Pe	Pectineus muscle
FA	Femoral artery	IP	Iliopsoas muscle	RFe	Rectus femoris muscle
FV	Femoral vein	IPA	Internal pudendal artery	Rt	Rectum
Fh	Femur, head	IUO	Internal urethral opening	SCo	Spermatic cord
GMT	Gluteus minimus tendon	IsS	Ischial spine	SGe	Superior gemellus muscle
GMa	Gluteus maximus muscle	LeA	Levator ani muscle	SN	Sciatic nerve
GMe	Gluteus medius muscle	ObA	Obturator artery	Sar	Sartorius muscle
GT	Greater trochanter	ObV	Obturator vein	SVe	Seminal vesicle

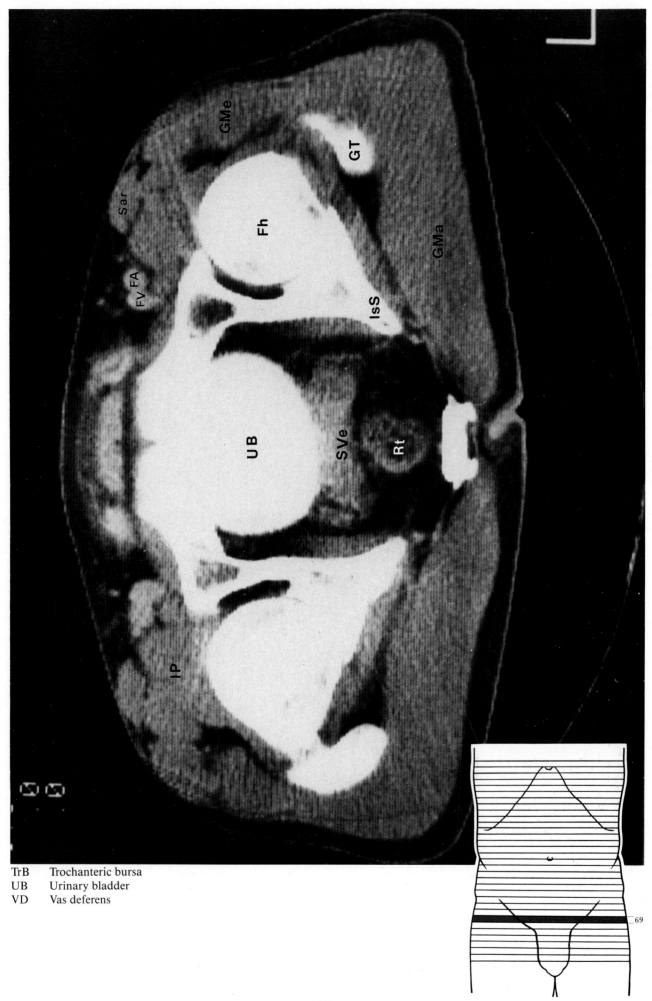

TrB   Trochanteric bursa
UB   Urinary bladder
VD   Vas deferens

Plate 70. TRANSVERSE PELVIS-MALE

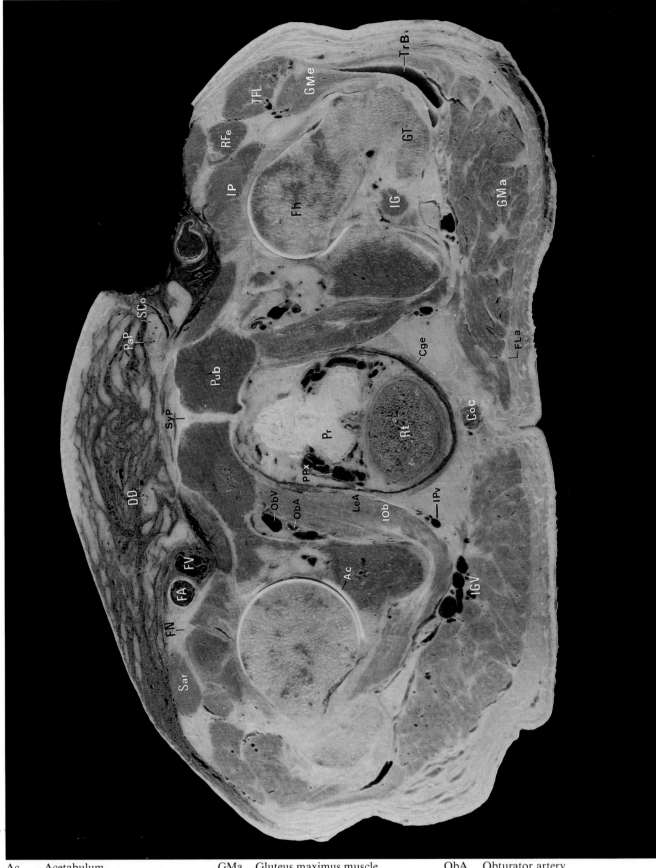

Ac	Acetabulum	GMa	Gluteus maximus muscle	ObA	Obturator artery
Cge	Coccygeus muscle	GMe	Gluteus medius muscle	ObV	Obturator vein
Coc	Coccyx	GT	Greater trochanter	PPx	Pudendal plexus
DD	Ductus deferens	IG	Inferior gemellus muscle	PaP	Panpiniform plexus
FA	Femoral artery	IGV	Inferior gluteal vein	Pr	Prostate
Fh	Femur, head	IOb	Internal obturator muscle	Pub	Pubic bone
FLa	Fascia lata	IP	Iliopsoas muscle	RFe	Rectus femoris muscle
FN	Femoral nerve	IPv	Internal pudendal vessel	Rt	Rectum
FV	Femoral vein	LeA	Levator ani muscle	SCo	Spermatic cord

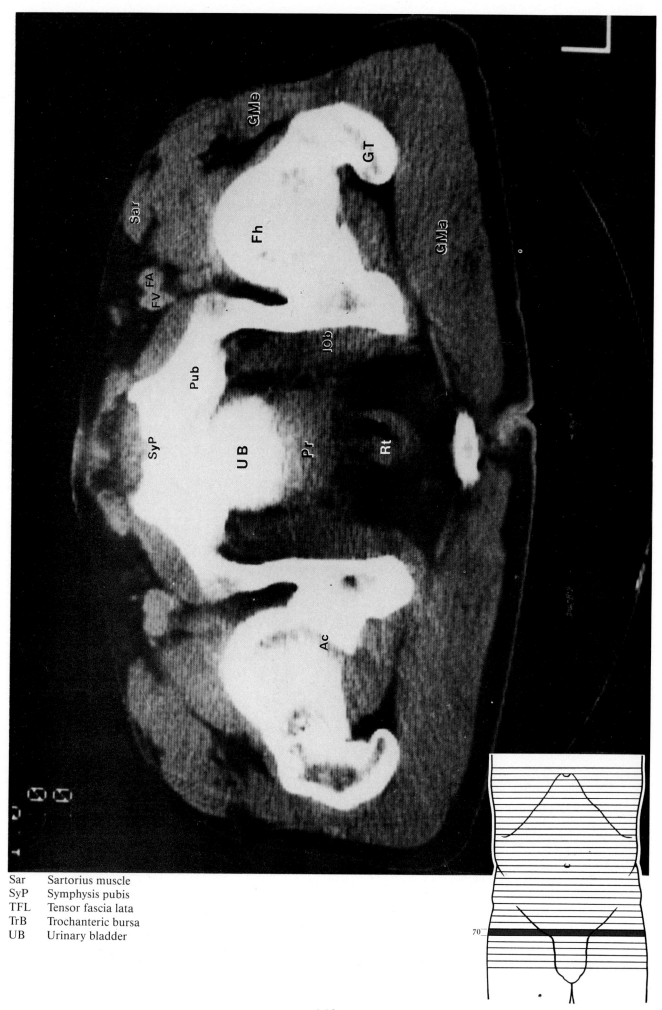

Sar	Sartorius muscle
SyP	Symphysis pubis
TFL	Tensor fascia lata
TrB	Trochanteric bursa
UB	Urinary bladder

Plate 71. TRANSVERSE PELVIS-MALE

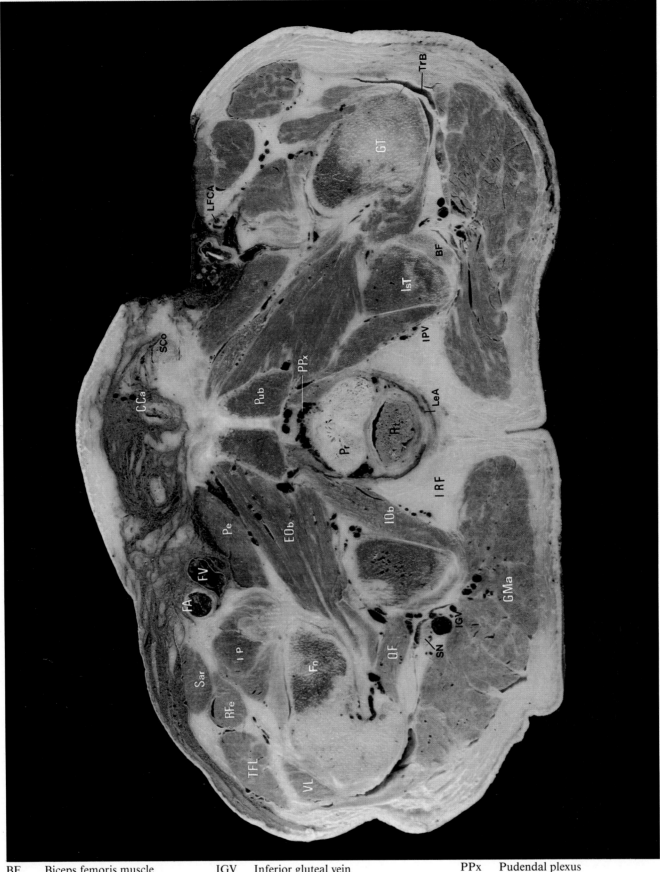

BF	Biceps femoris muscle	IGV	Inferior gluteal vein	PPx	Pudendal plexus
CCa	Corpus cavernosum	IOb	Internal obturator muscle	Pe	Pectineus muscle
Coc	Coccyx	IP	Iliopsoas muscle	Pr	Prostate
EOb	External obturator muscle	IPv	Internal pudendal vessel	Pub	Pubic bone
FA	Femoral artery	IRF	Ischiorectal fossa	RFe	Rectus femoris muscle
FV	Femoral vein	IsT	Ischial tuberosity	Rt	Rectum
Fn	Femur, neck	LFCA	Lateral femoral circumflex artery	SCo	Spermatic cord
GMa	Gluteus maximus muscle	LeA	Levator ani muscle	SN	Sciatic nerve
GT	Greater trochanter	OF	Omental fat	Sar	Sartorius muscle

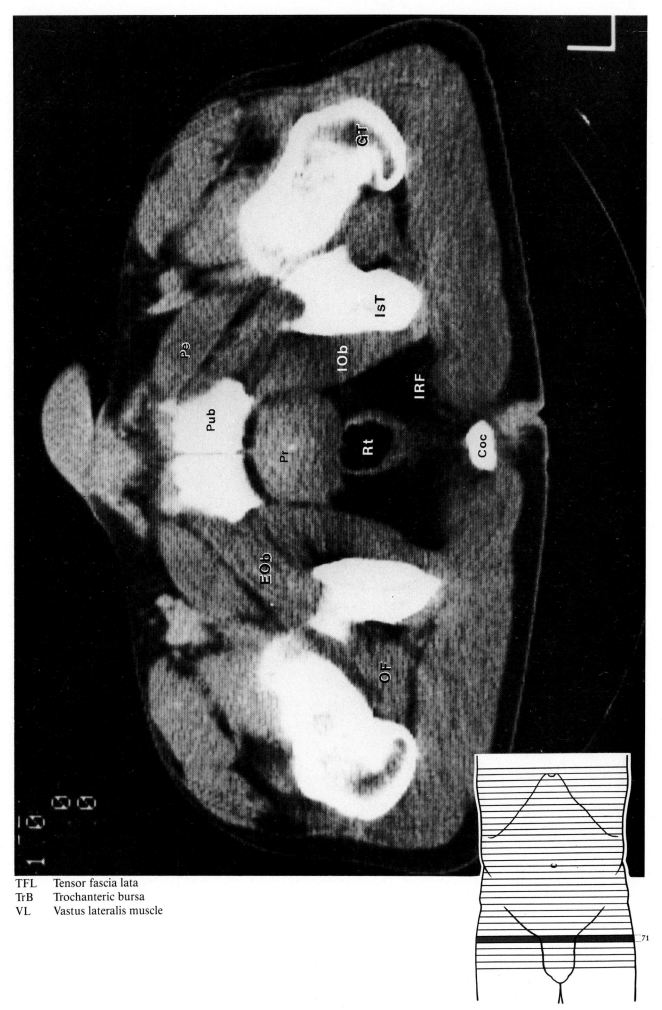

TFL   Tensor fascia lata
TrB   Trochanteric bursa
VL   Vastus lateralis muscle

Plate 72. TRANSVERSE PELVIS-MALE

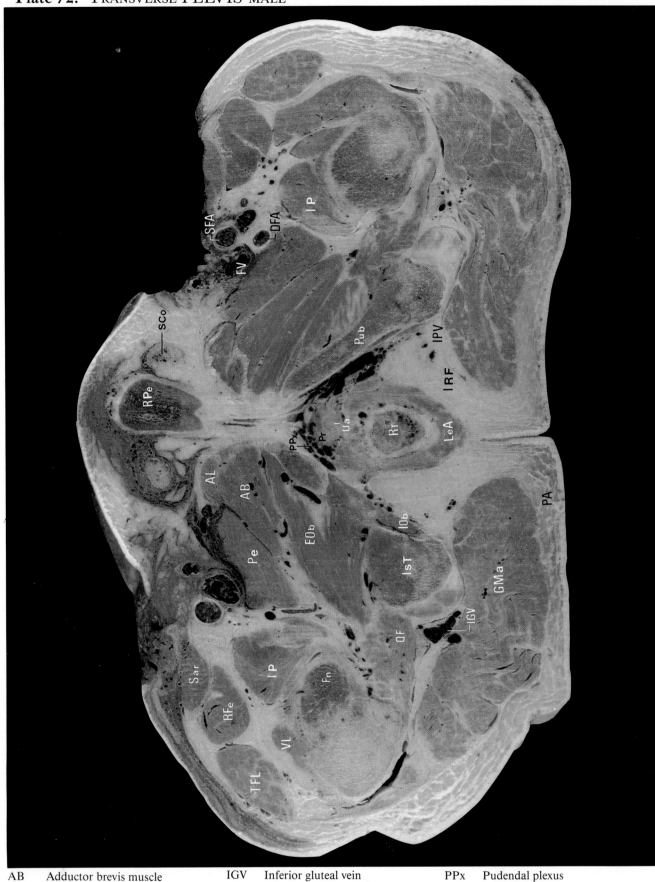

AB	Adductor brevis muscle	IGV	Inferior gluteal vein
AL	Adductor longus muscle	IOb	Internal obturator muscle
Coc	Coccyx	IP	Iliopsoas muscle
DFA	Deep femoral artery	IPv	Internal pudendal vessel
EOb	External obturator muscle	IRF	Ischiorectal fossa
FV	Femoral vein	IsT	Ischial tuberosity
Fn	Femur, neck	LeA	Levator ani muscle
GMa	Gluteus maximus muscle	OF	Omental fat
GT	Greater trochanter	PA	Panniculus adiposus

PPx	Pudendal plexus
Pe	Pectineus muscle
Pr	Prostate
Pub	Pubic bone
RFe	Rectus femoris muscle
RPe	Root of penis
Rt	Rectum
SCo	Spermatic cord
SFA	Superficial femoral artery

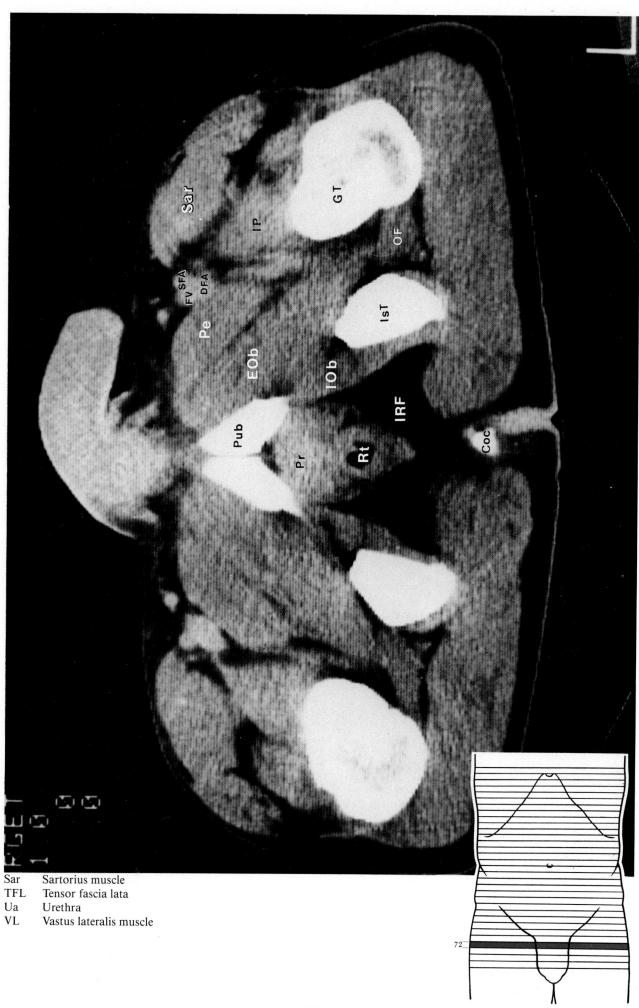

Sar   Sartorius muscle
TFL   Tensor fascia lata
Ua   Urethra
VL   Vastus lateralis muscle

Plate 73. TRANSVERSE PELVIS-MALE

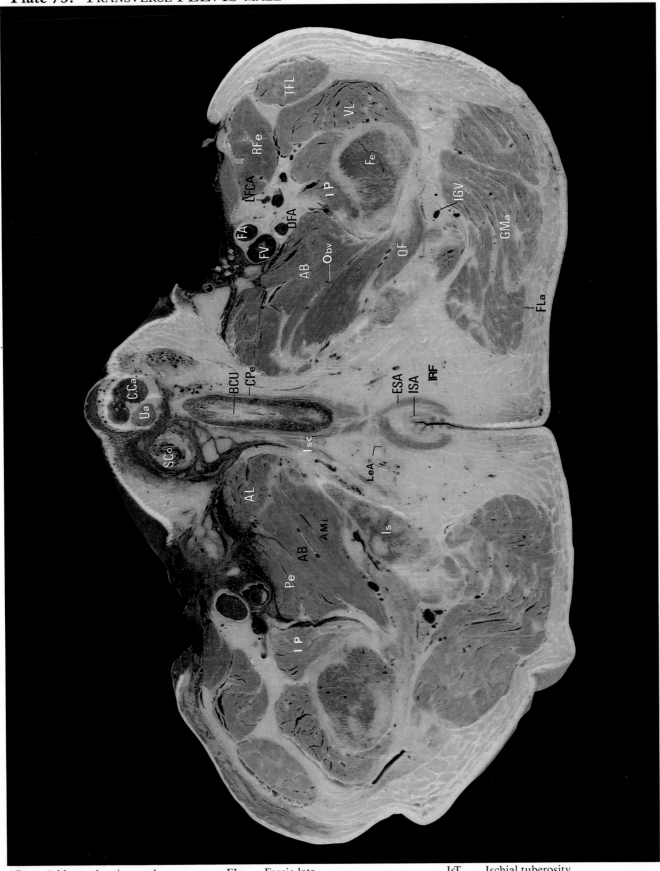

AB	Adductor brevis muscle	FLa	Fascia lata	IsT	Ischial tuberosity
AL	Adductor longus muscle	FV	Femoral vein	Isc	Ischiocavernous muscle
AMi	Adductor minimus muscle	Fe	Femur	LFCA	Lateral femoral circumflex artery
BCU	Bullocavernous urethra	GMa	Gluteus maximus muscle	LeA	Levator ani muscle
CCa	Corpus cavernosum	IGV	Inferior gluteal vein	OF	Omental fat
CPe	Crus penis	IP	Iliopsoas muscle	Obv	Obsturator vessel
DFA	Deep femoral artery	IRF	Ischiorectal fossa	Pe	Pectineus muscle
ESA	External sphincter ani	ISA	Internal sphincter ani	Pr	Prostate
FA	Femoral artery	Is	Ischium	RFe	Rectus femoris muscle

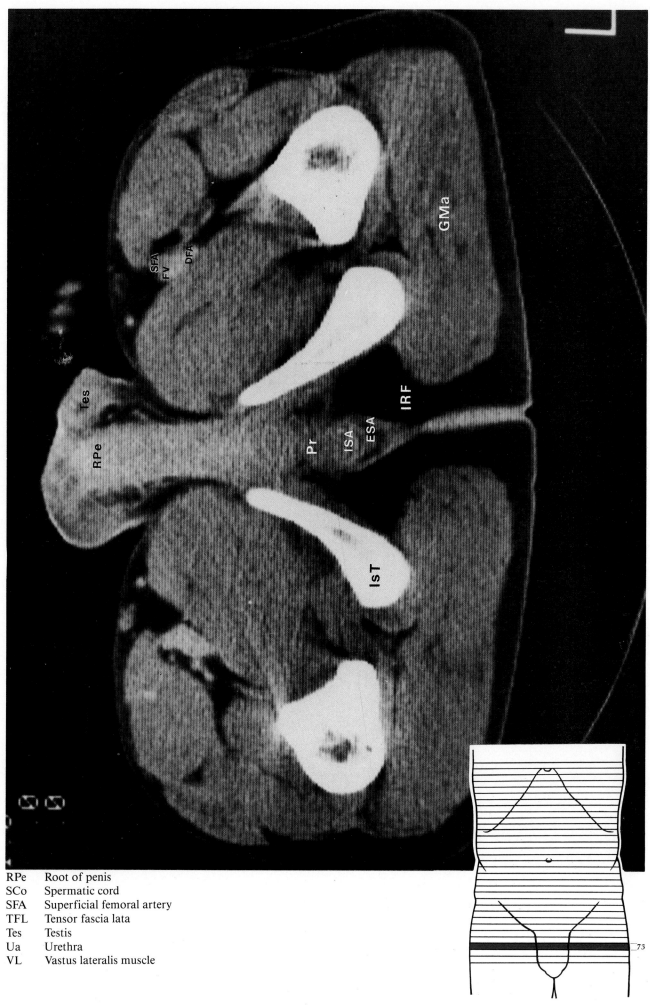

RPe	Root of penis
SCo	Spermatic cord
SFA	Superficial femoral artery
TFL	Tensor fascia lata
Tes	Testis
Ua	Urethra
VL	Vastus lateralis muscle

Plate 74. TRANSVERSE PELVIS-MALE

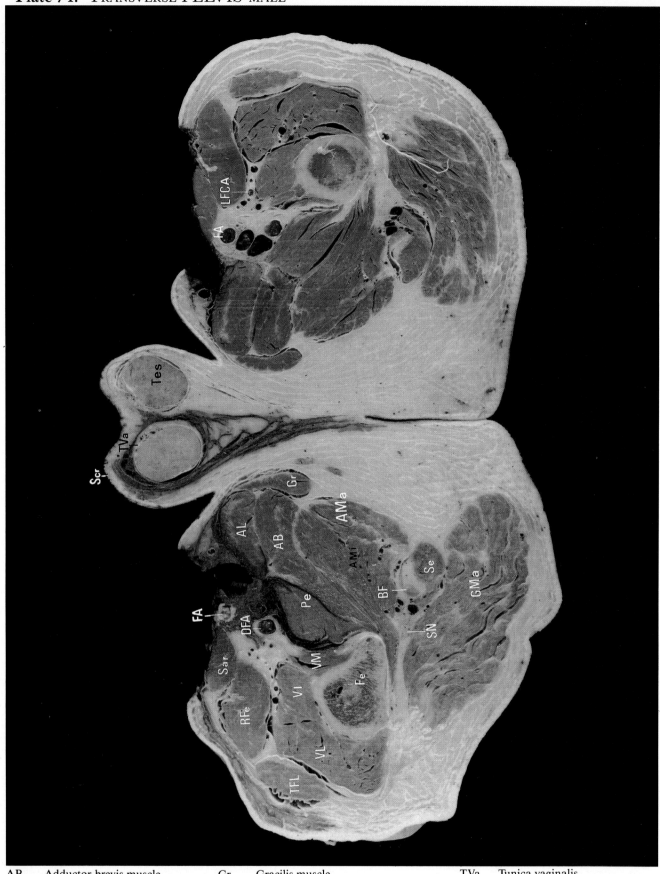

AB	Adductor brevis muscle	Gr	Gracilis muscle	TVa	Tunica vaginalis
AL	Adductor longus muscle	LFCA	Lateral femoral circumflex artery	Tes	Testis
AMa	Adductor magnus muscle	Pe	Pectineus muscle	VI	Vastus intermedius muscle
AMi	Adductor minimus muscle	RFe	Rectus femoris muscle	VL	Vastus lateralis muscle
BF	Biceps femoris muscle	SN	Sciatic nerve	VM	Vastus medialis muscle
DFA	Deep femoral artery	Sar	Sartorius muscle		
FA	Femoral artery	Scr	Scrotum		
Fe	Femur	Se	Semitendinosus muscle		
GMa	Gluteus maximus muscle	TFL	Tensor fascia lata		

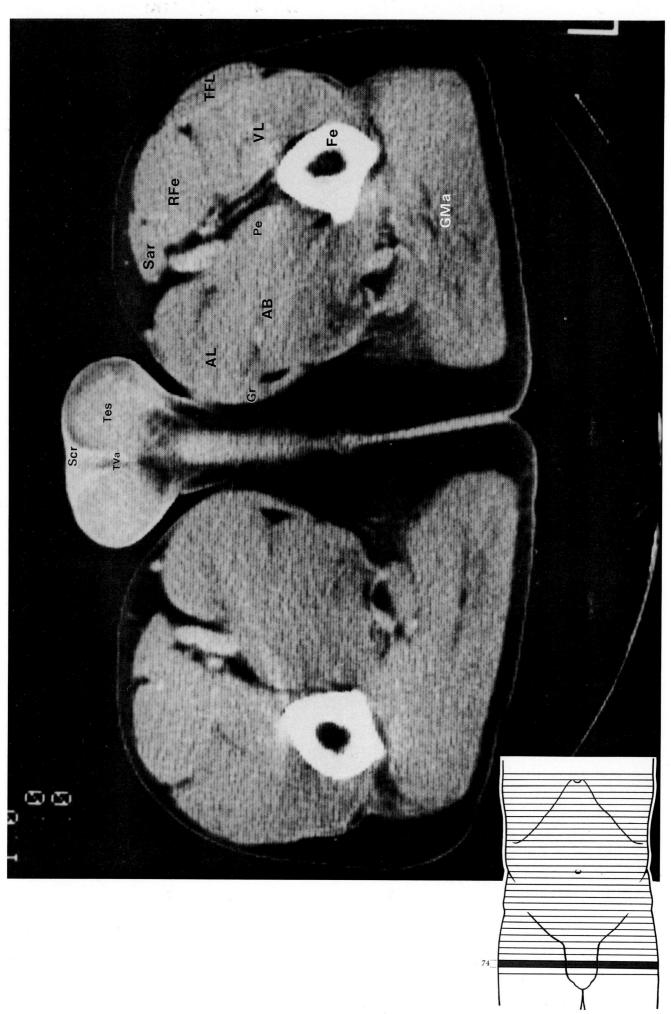

Plate 75. TRANSVERSE PELVIS-MALE

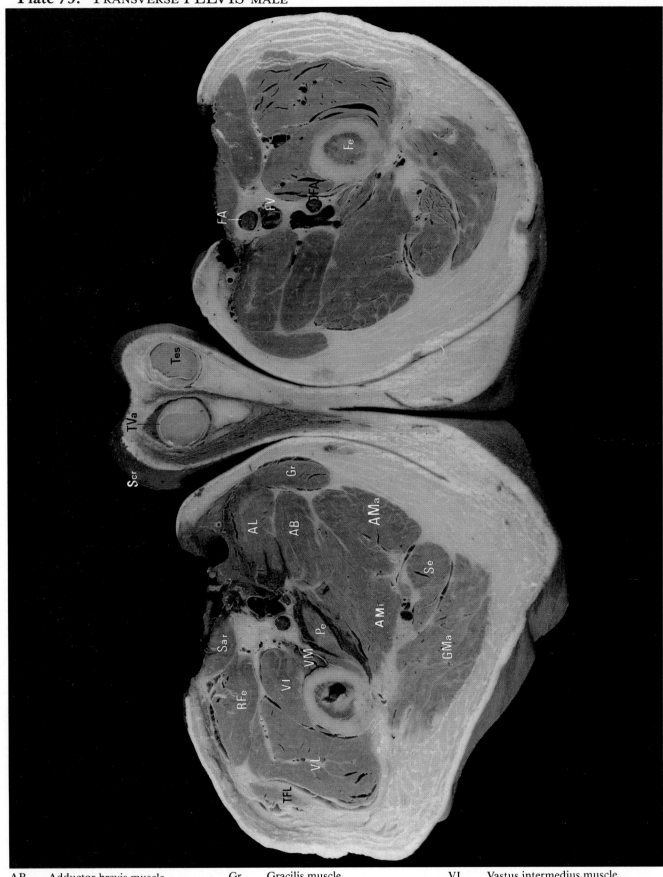

AB	Adductor brevis muscle	Gr	Gracilis muscle	VI	Vastus intermedius muscle
AL	Adductor longus muscle	Pe	Pectineus muscle	VM	Vastus medialis muscle
AMa	Adductor magnus muscle	RFe	Rectus femoris muscle		
AMi	Adductor minimus muscle	Sar	Sartorius muscle		
DFA	Deep femoral artery	Scr	Scrotum		
FA	Femoral artery	Se	Semitendinosus muscle		
FV	Femoral vein	TFL	Tensor fascia lata		
Fe	Femur	TVa	Tunica vaginalis		
GMa	Gluteus maximus muscle	Tes	Testis		

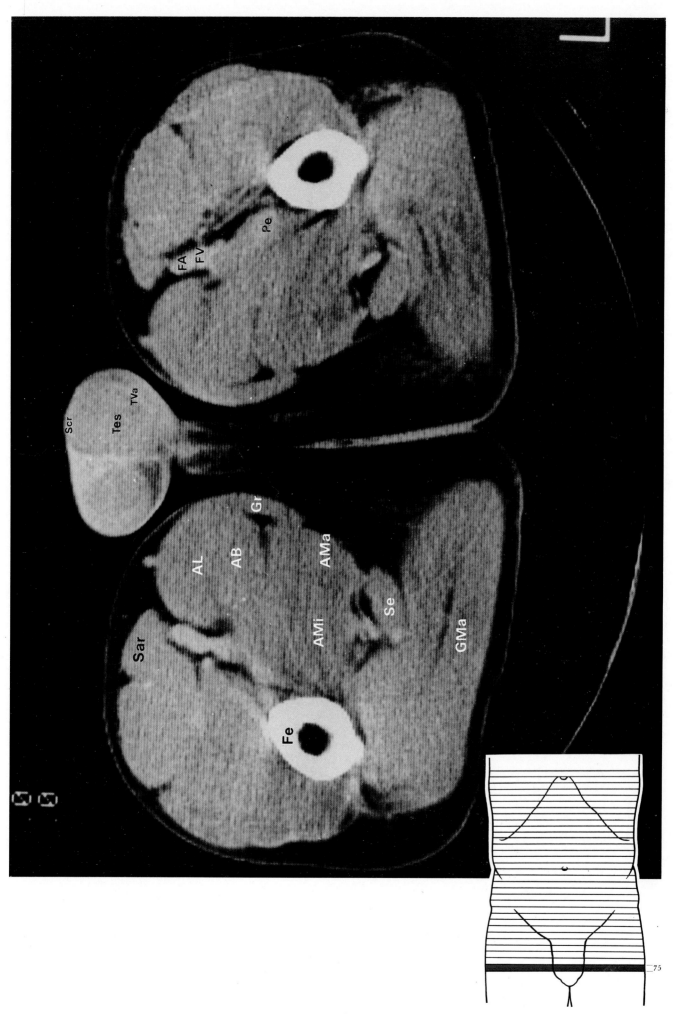

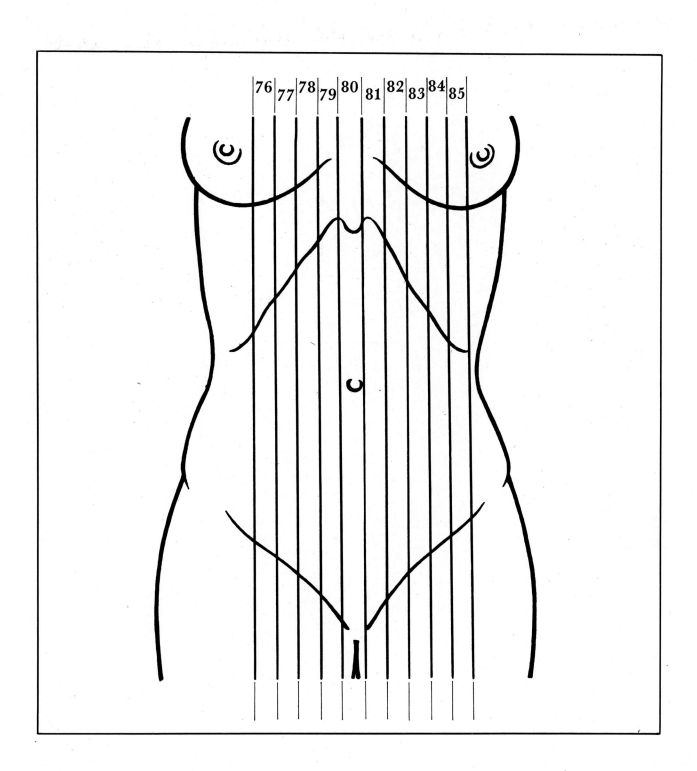

Sagittal Section-female (Specimen) Plate 76-85

Plate 76. Sᴀɢɪᴛᴛᴀʟ ABDOMEN ᴀɴᴅ PELVIS-ꜰᴇᴍᴀʟᴇ

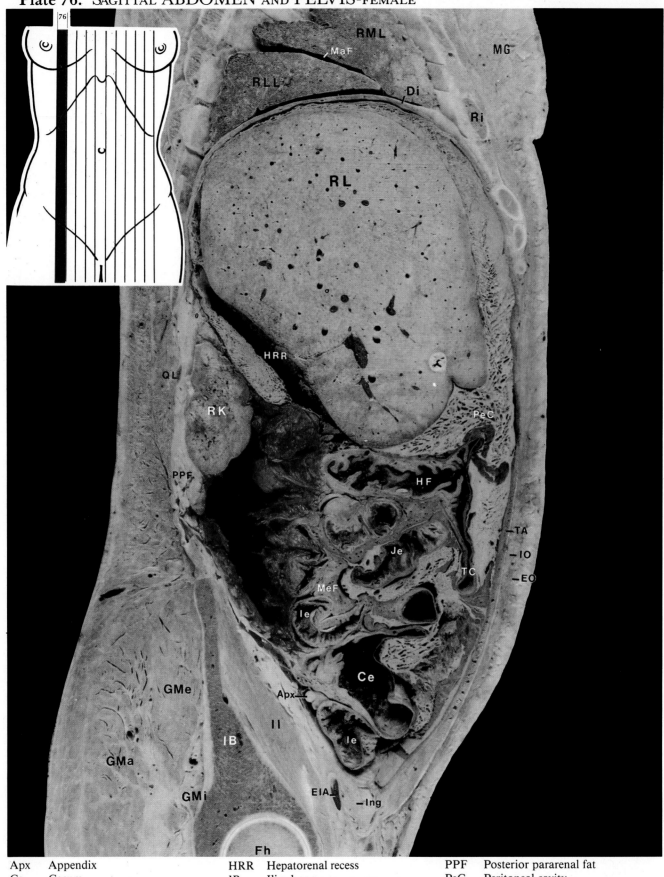

Apx	Appendix	HRR	Hepatorenal recess	PPF	Posterior pararenal fat
Ce	Cecum	IB	Iliac bone	PeC	Peritoneal cavity
Di	Diaphragm	IO	Internal oblique muscle	QL	Quadratus lumborum muscle
EIA	External iliac artery	Ie	Ileum	RK	Right kidney
EO	External oblique muscle	Il	Iliacus muscle	RL	Right lobe of liver
Fh	Femur, head	Ing	Inguinal ligament	RLL	Right lower lobe of lung
GMa	Gluteus maximus muscle	Je	Jejunum	RML	Right middle lobe of lung
GMe	Gluteus medius muscle	MG	Mammary gland	Ri	Rib
GMi	Gluteus minimus muscle	MaF	Major fissure	TA	Transversus abdominis muscle
HF	Hepatic flexure of colon	MeF	Mesenteric fat	TC	Transverse colon

Plate 77. SAGITTAL ABDOMEN AND PELVIS-FEMALE

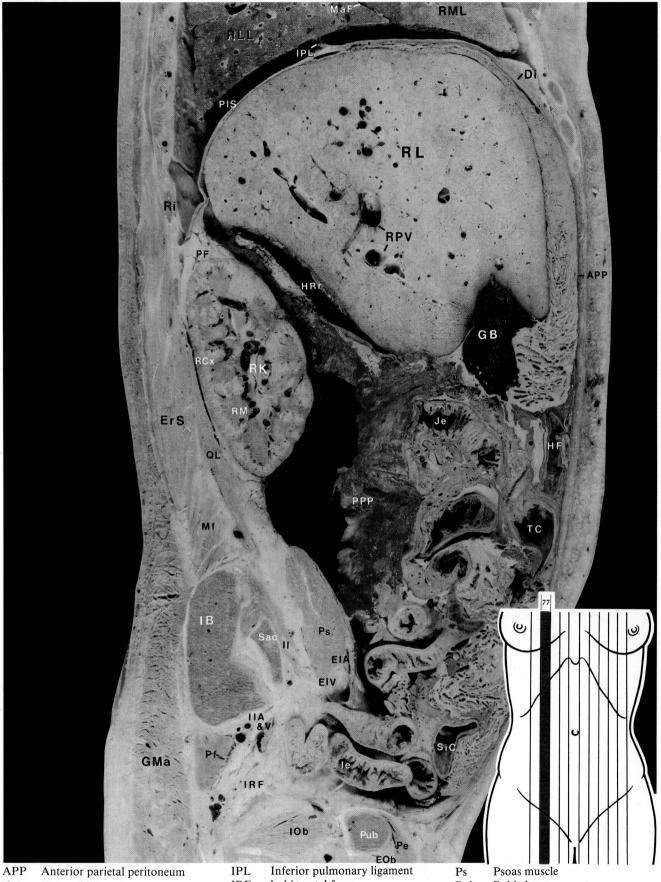

APP	Anterior parietal peritoneum	IPL	Inferior pulmonary ligament	Ps	Psoas muscle
EIA	External iliac artery	IRF	Ischiorectal fossa	Pub	Pubic bone
EIV	External iliac vein	Il	Iliacus muscle	QL	Quadratus lumborum muscle
ErS	Erector spinae muscle	Mf	Multifidus muscle	RCx	Renal cortex
GB	Gall bladder	PF	Perirenal fat	RM	Renal medulla
HRR	Hepatorenal recess	PPP	Posterior parietal peritoneum	RPV	Right portal vein
IIA	Internal iliac artery	Pe	Pectineus muscle	Sac	Sacrum
IIV	Internal iliac vein	Pf	Piriformis muscle	SiC	Sigmoid colon
IOb	Internal obturator muscle	PlS	Pleural space		

Plate 78. SAGITTAL ABDOMEN AND PELVIS-FEMALE

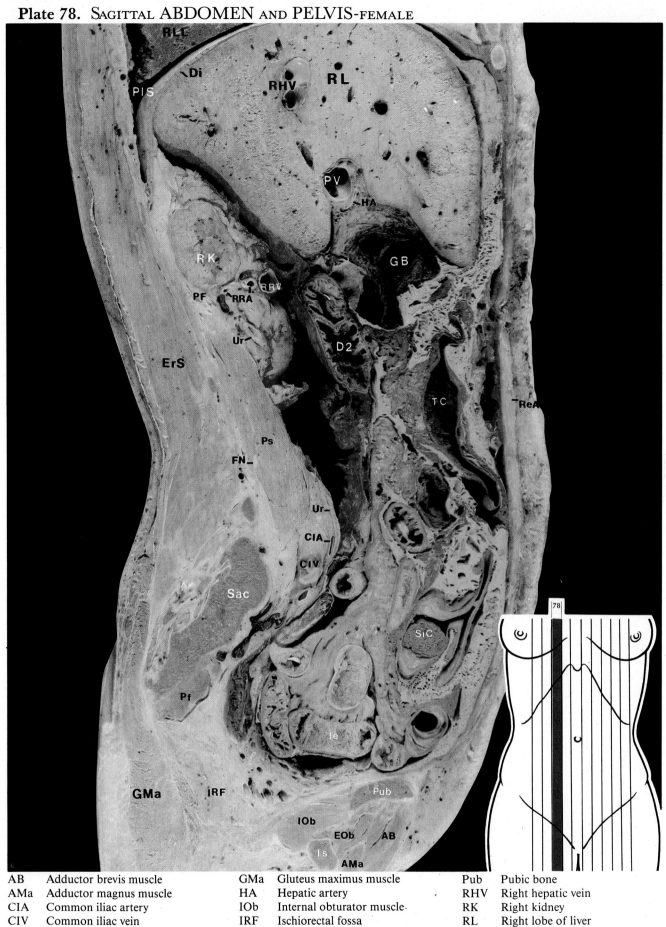

AB	Adductor brevis muscle	GMa	Gluteus maximus muscle	Pub	Pubic bone
AMa	Adductor magnus muscle	HA	Hepatic artery	RHV	Right hepatic vein
CIA	Common iliac artery	IOb	Internal obturator muscle	RK	Right kidney
CIV	Common iliac vein	IRF	Ischiorectal fossa	RL	Right lobe of liver
D2	Second portion of duodenum	Ie	Ileum	RRA	Right renal artery
Di	Diaphragm	PF	Perirenal fat	RRV	Right renal vein
EOb	External obturator muscle	PV	portal vein	ReA	Rectus abdominis muscle
ErS	Erector spinae muscle	Pf	Pyriformis muscle	Sac	Sacrum
FN	Femoral nerve	PlS	Pleural space	Ur	Ureter
GB	Gall bladder	Ps	Psoas muscle		

Plate 79. SAGITTAL ABDOMEN AND PELVIS-FEMALE

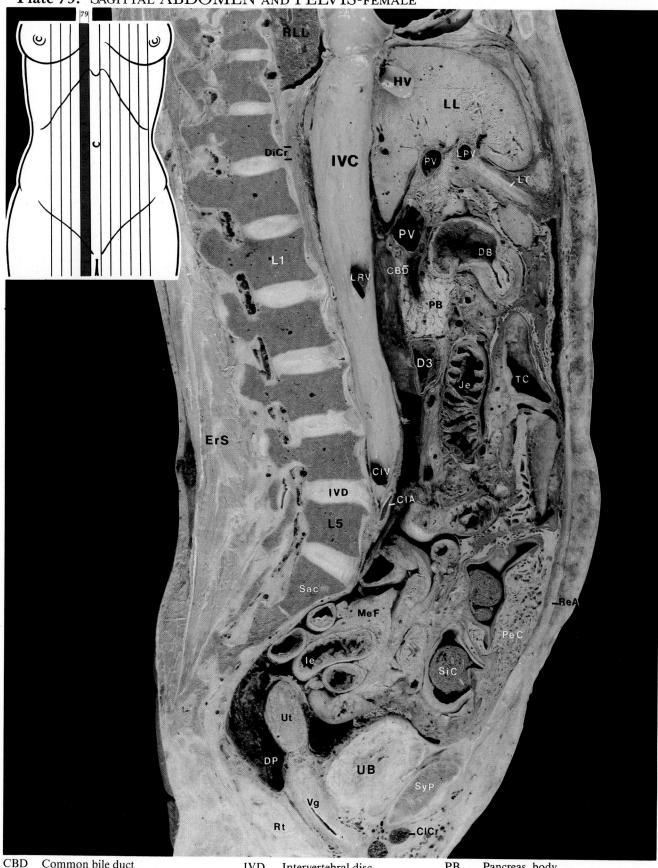

CBD	Common bile duct	IVD	Intervertebral disc	PB	Pancreas, body
CIA	Common iliac artery	Ie	Ileum	PeC	Peritoneal cavity
CIV	Common iliac vein	Je	Jejunum	RLL	Right lower lobe of lung
ClCr	Clitoris, crus	L1	L1 vertebral body	Rt	Rectum
D3	Third portion of duodenum	L5	L5 vertebral body	SiC	Sigmoid colon
DB	Duodenal bulb	LL	Left lobe of liver	SyP	Symphysis pubis
DP	Douglas pouch	LPV	Left portal vein	TC	Transverse colon
DiCr	Diaphragmatic crus	LRV	Left renal vein	UB	Urinary bladder
HV	Hepatic vein	LT	Ligamentum teres	Ut	Uterus
IVC	Inferior vena cava	MeF	Mesenteric fat	Vg	Vagina

Plate 80. SAGITTAL ABDOMEN AND PELVIS-FEMALE

Ao	Aorta	HA	Hepatic artery	PB	Pancreas, body
CA	Celiac axis	IVD	Intervertebral disc	PV	Portal vein
CIA	Common iliac artery	Ie	Ileum	Rt	Rectum
CIV	Common iliac vein	Je	Jejunum	SMA	Superior mesenteric artery
CL	Caudate lobe of liver	L5	L5 vertebral body	SiC	Sigmoid colon
D4	Fourth portion of duodenum	LHV	Left hepatic vein	StA	Stomach, antrum
DiCr	Diaphragmatic crus	LL	Left lobe of liver	SyP	Symphysis pubis
ErS	Erector spinae muscle	LOt	Lesser omentum	TC	Transverse colon
Es	Esophagus	LRV	Left renal vein	UB	Urinary bladder
GOt	Greater omentum	LSa	Lesser sac	Ut	Uterus

Plate 81. SAGITTAL ABDOMEN AND PELVIS-FEMALE

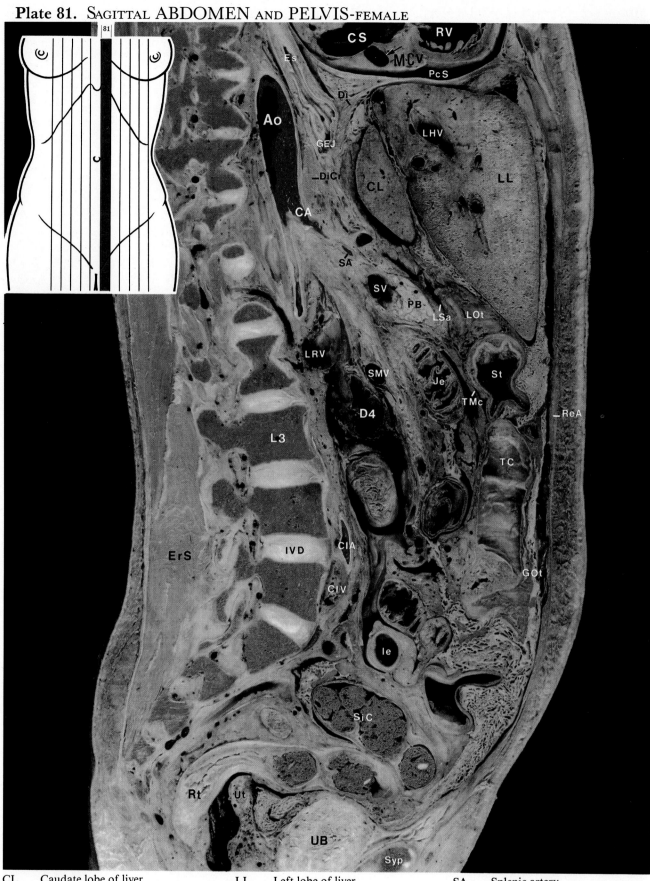

CL	Caudate lobe of liver	LL	Left lobe of liver	SA	Splenic artery
CS	Coronary sinus	LOt	Lesser omentum	SMV	Superior mesenteric vein
D4	Fourth portion of duodenum	LRV	Left renal vein	SV	Splenic vein
Di	Diaphragm	LSa	Lesser sac	SiC	Sigmoid colon
DiCr	Diaphragmatic crus	MCV	Middle cardiac vein	St	Stomach
ErS	Erector spinae muscle	PB	Pancreas, body	SyP	Symphysis pubis
GEJ	Gastroesophageal junction	PcS	Pericardial space	TC	Transverse colon
GOt	Greater omentum	RV	Right ventricle of heart	TMc	Transverse mesocolon
L3	L3 vertebral body	ReA	Rectus abdominis muscle	UB	Urinary bladder
LHV	Left hepatic vein	Rt	Rectum	Ut	Uterus

Plate 82. SAGITTAL ABDOMEN AND PELVIS-FEMALE

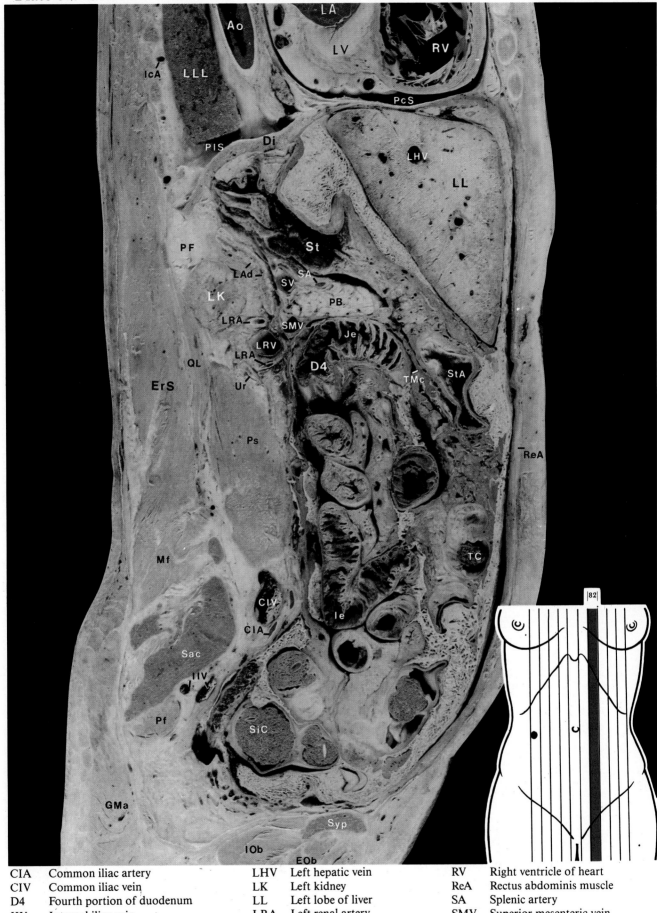

CIA	Common iliac artery	LHV	Left hepatic vein	RV	Right ventricle of heart
CIV	Common iliac vein	LK	Left kidney	ReA	Rectus abdominis muscle
D4	Fourth portion of duodenum	LL	Left lobe of liver	SA	Splenic artery
IIV	Internal iliac vein	LRA	Left renal artery	SMV	Superior mesenteric vein
IOb	Internal obturator muscle	LRV	Left renal vein	SV	Splenic vein
IcA	Intercostal artery	Mf	Multifidus muscle	Sac	Sacrum
Ie	Ileum	PB	Pancreas, body	StA	Stomach, antrum
Je	Jejunum	PF	Perirenal fat	SyP	Symphysis pubis
LAd	Left adrenal gland	Pf	Pyriformis muscle	Ur	Ureter

Plate 83. SAGITTAL ABDOMEN AND PELVIS-FEMALE

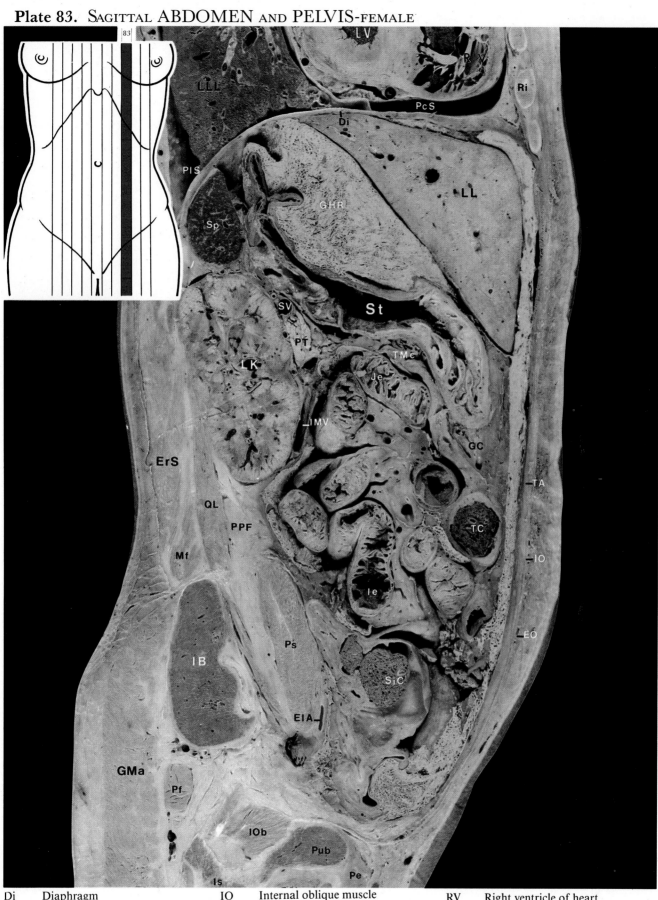

Di	Diaphragm	IO	Internal oblique muscle	RV	Right ventricle of heart
EIA	External iliac artery	LK	Left kidney	Ri	Rib
EO	External oblique muscle	Mf	Multifidus muscle	SV	Splenic vein
ErS	Erector spinae muscle	PPF	Posterior pararenal fat	SiC	Sigmoid colon
GC	Gastrocolic ligament	PT	Pancreas, tail	Sp	Spinalis muscle
GHR	Gastrohepatic recess	Pe	Pectineus muscle	St	Stomach
GMa	Gluteus maximus muscle	Ps	Psoas muscle	TA	Transversus abdominis muscle
IB	Iliac bone	Pub	Pubic bone	TC	Transverse colon
IMv	Internal mammary vessel	QL	Quadratus lumborum muscle	TMc	Transverse mesocolon

Plate 84. SAGITTAL ABDOMEN AND PELVIS-FEMALE

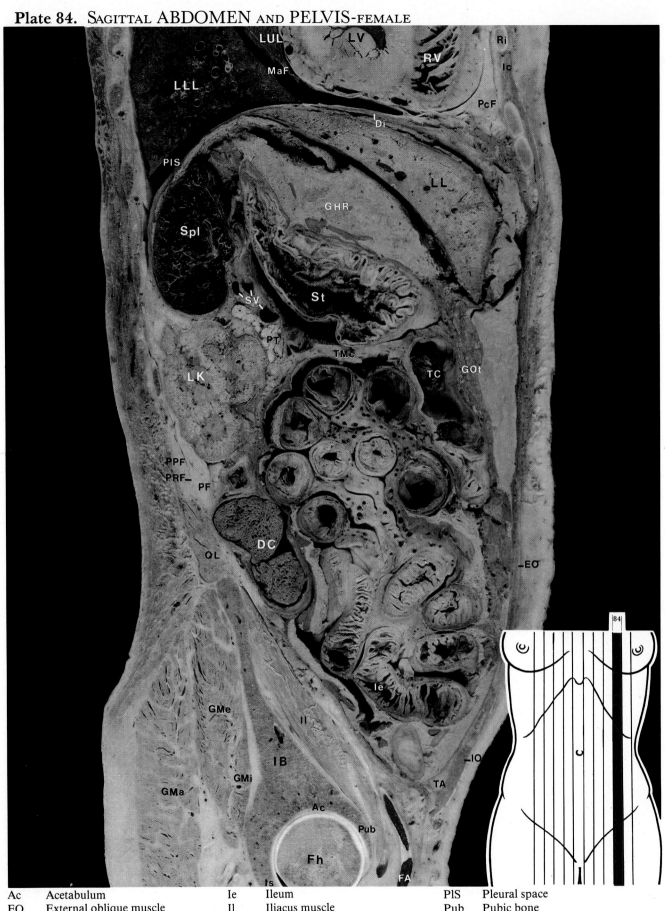

Ac	Acetabulum	Ie	Ileum	PlS	Pleural space
EO	External oblique muscle	Il	Iliacus muscle	Pub	Pubic bone
FA	Femoral artery	Is	Ischium	QL	Quadratus lumborum muscle
Fh	Femur, head	LK	Left kidney	RV	Rightr ventricle of heart
GHR	Gastrohepatic recess	MaF	Major fissure	Spl	Spleen
GOt	Greater omentum	PF	Perirenal fat	St	Stomach
IB	Iliac bone	PPF	Posterior pararenal fat	TA	Transversus abdominis muscle
IO	Internal oblique muscle	PRF	Posterior renal fascia	TC	Transverse colon
Ic	Intercostal muscle	PT	Pancreas, tail	TMc	Transverse mesocolon

Plate 85. SAGITTAL ABDOMEN AND PELVIS-FEMALE

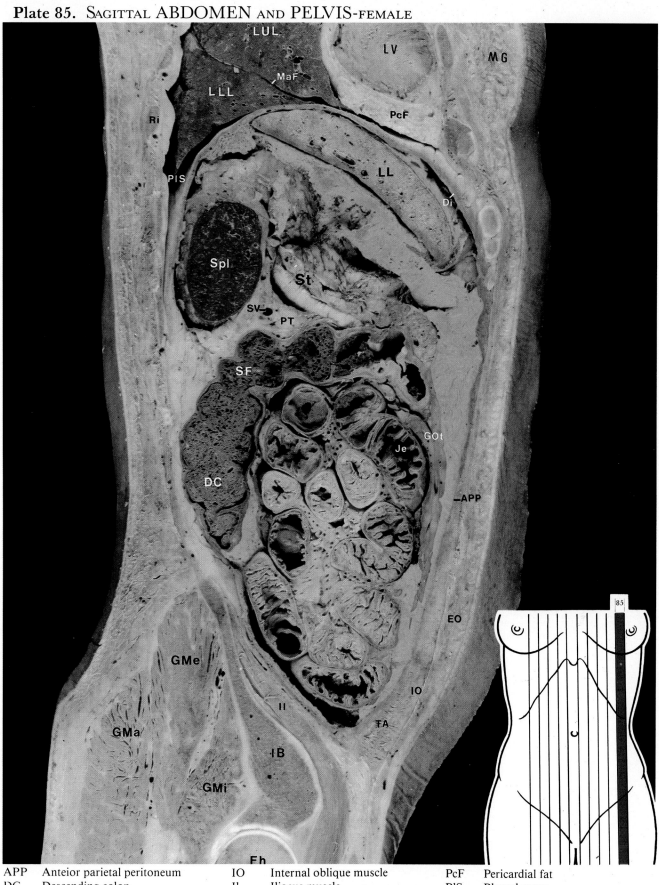

APP	Anteior parietal peritoneum	IO	Internal oblique muscle	PcF	Pericardial fat
DC	Descending colon	Il	Iliacus muscle	PlS	Pleural space
Di	Diaphragm	Je	Jejunum	Ri	Rib
EO	External oblique muscle	LL	Left lobe of liver	SF	Splenic flexure
Fh	Femur, head	LLL	Left lower lobe of lung	SV	Splenic vein
GMa	Gluteus maximus muscle	LUL	Left upper lobe of lung	Spl	Spleen
GMe	Gluteus medius muscle	LV	Left ventricle of heart	St	Stomach
GMi	Gluteus minimus muscle	MG	Mammary gland	TA	Transversus abdominis muscle
GOt	Greater omentum	MaF	Major fissure		
IB	Iliac bone	PT	Pancreas, tail		

PART 4
MAGNETIC RESONANCE IMAGING

BRAIN, HEAD AND NECK

CHEST

ABDOMEN AND PELVIS

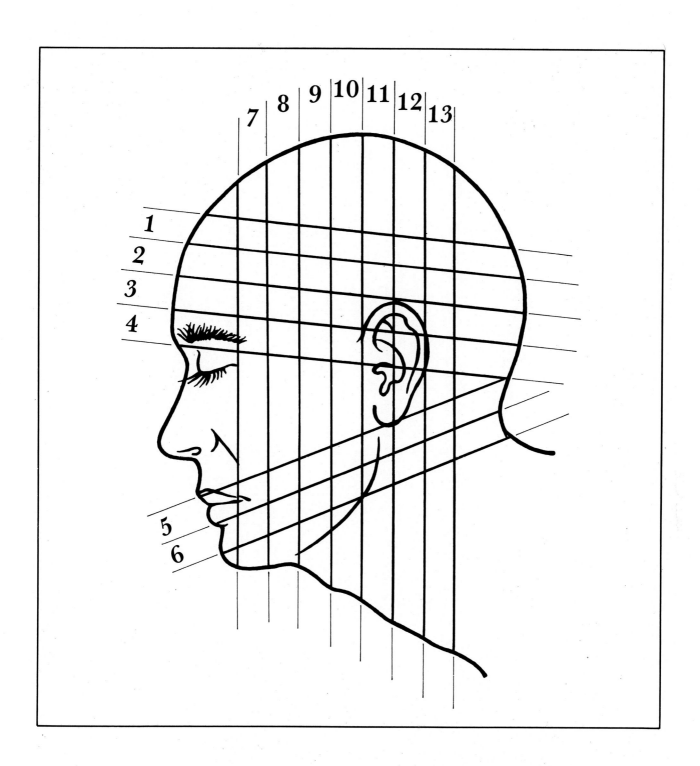

BRAIN, HEAD AND NECK   Transverse MRI 1-6
                        Coronal MRI 7-13

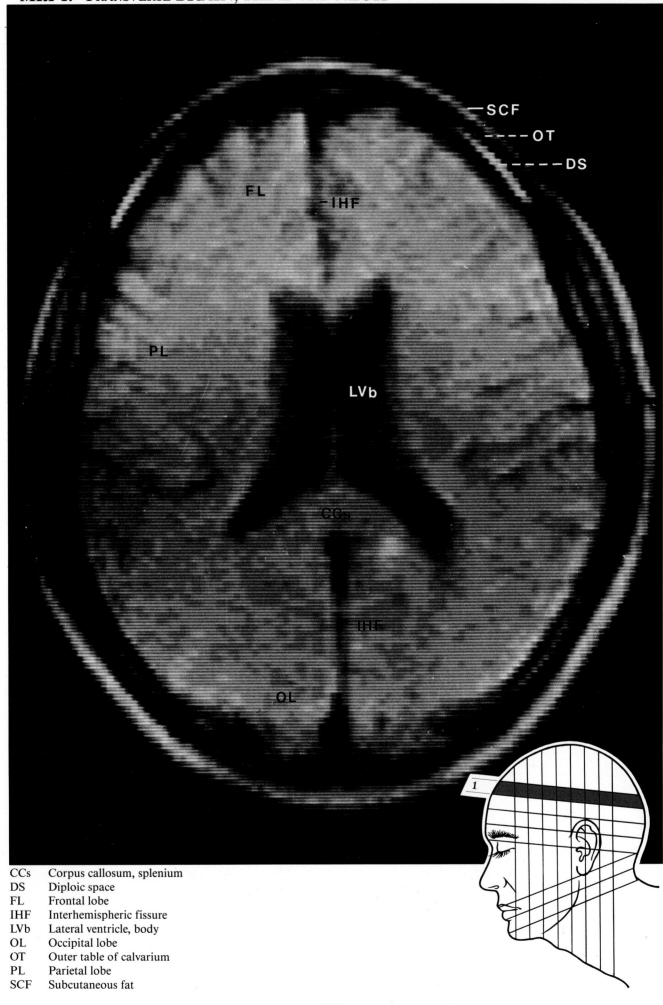

CCs	Corpus callosum, splenium
DS	Diploic space
FL	Frontal lobe
IHF	Interhemispheric fissure
LVb	Lateral ventricle, body
OL	Occipital lobe
OT	Outer table of calvarium
PL	Parietal lobe
SCF	Subcutaneous fat

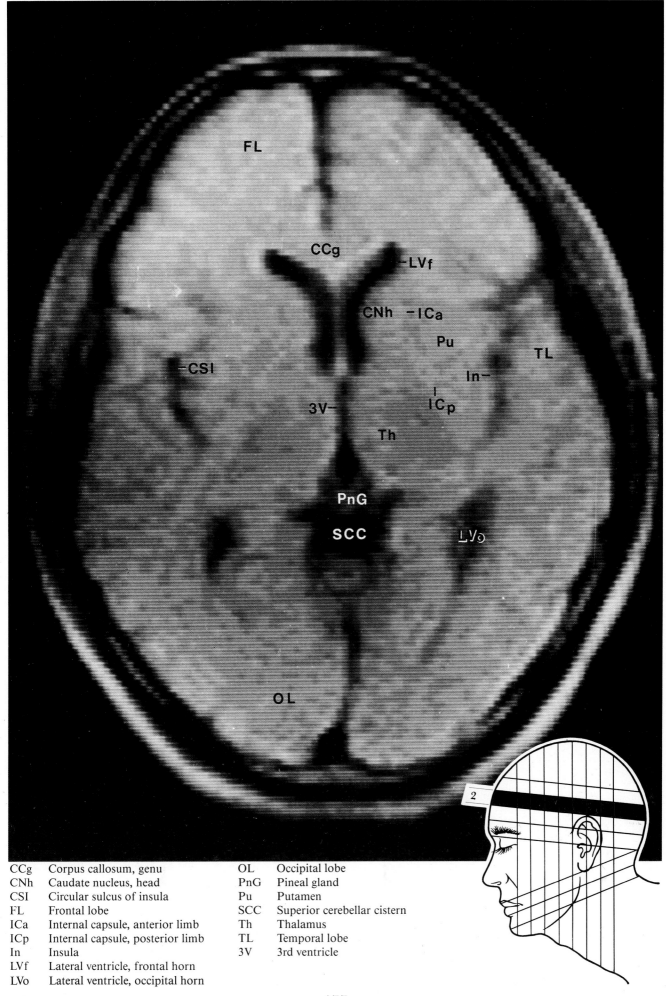

CCg	Corpus callosum, genu	OL	Occipital lobe
CNh	Caudate nucleus, head	PnG	Pineal gland
CSI	Circular sulcus of insula	Pu	Putamen
FL	Frontal lobe	SCC	Superior cerebellar cistern
ICa	Internal capsule, anterior limb	Th	Thalamus
ICp	Internal capsule, posterior limb	TL	Temporal lobe
In	Insula	3V	3rd ventricle
LVf	Lateral ventricle, frontal horn		
LVo	Lateral ventricle, occipital horn		

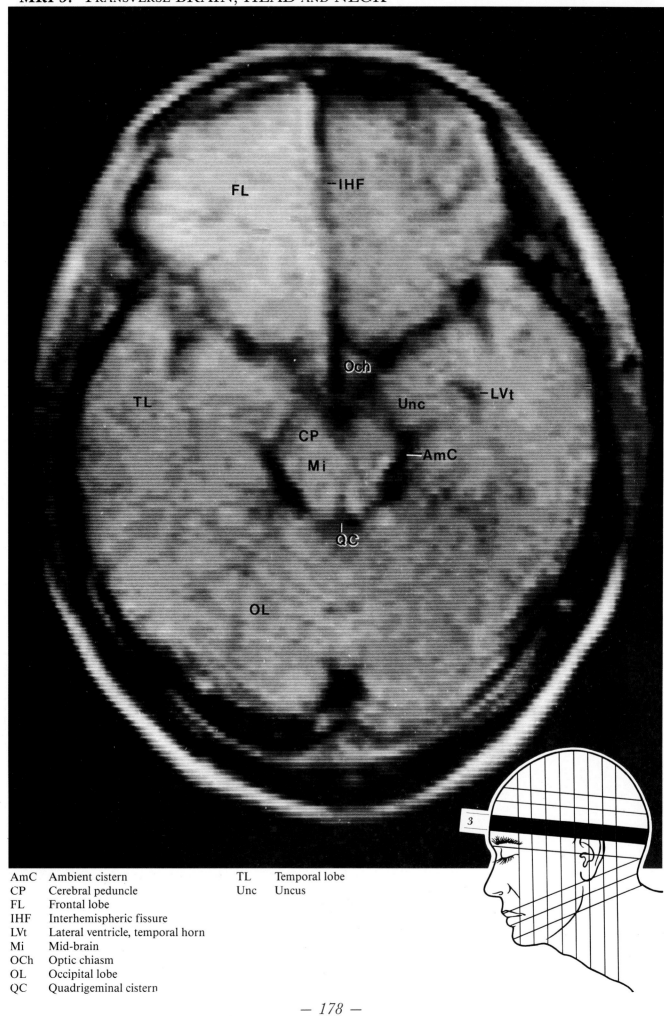

AmC   Ambient cistern
CP   Cerebral peduncle
FL   Frontal lobe
IHF   Interhemispheric fissure
LVt   Lateral ventricle, temporal horn
Mi   Mid-brain
OCh   Optic chiasm
OL   Occipital lobe
QC   Quadrigeminal cistern

TL   Temporal lobe
Unc   Uncus

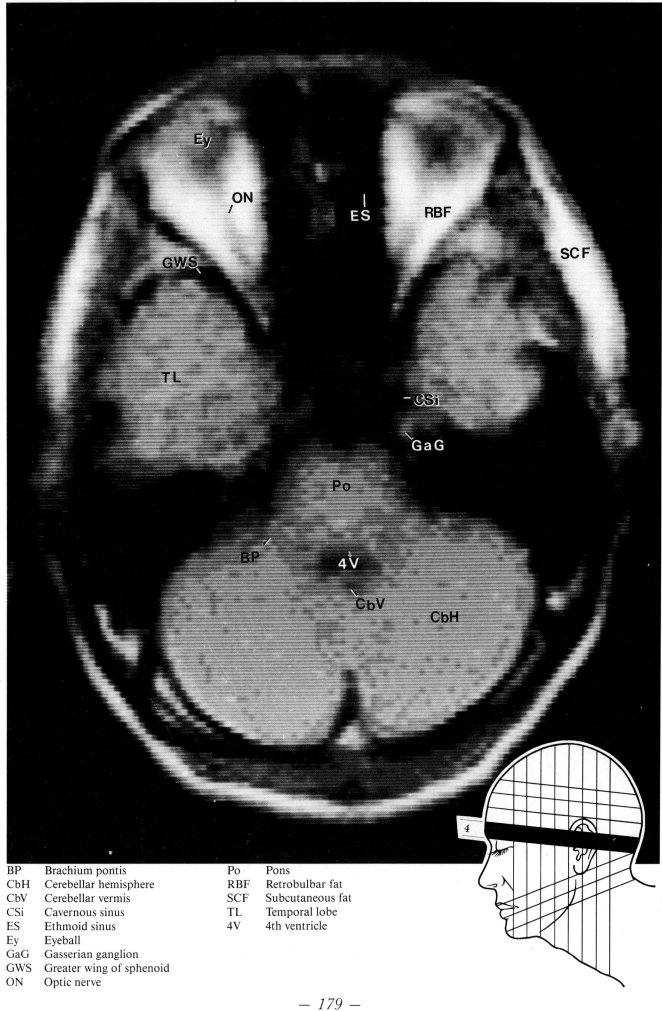

BP	Brachium pontis	Po	Pons
CbH	Cerebellar hemisphere	RBF	Retrobulbar fat
CbV	Cerebellar vermis	SCF	Subcutaneous fat
CSi	Cavernous sinus	TL	Temporal lobe
ES	Ethmoid sinus	4V	4th ventricle
Ey	Eyeball		
GaG	Gasserian ganglion		
GWS	Greater wing of sphenoid		
ON	Optic nerve		

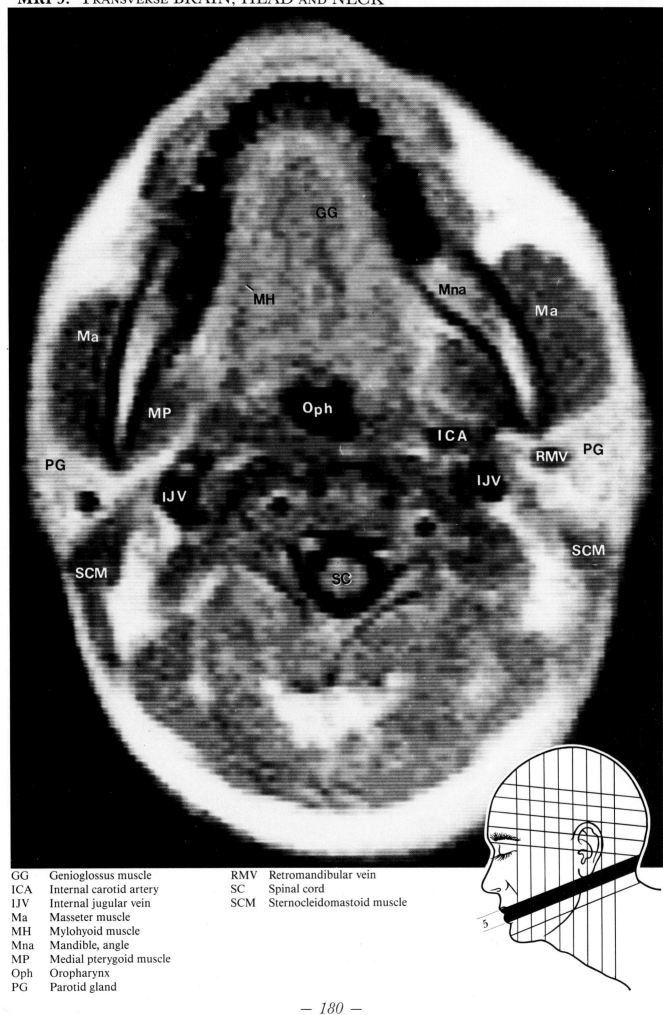

GG	Genioglossus muscle	RMV	Retromandibular vein
ICA	Internal carotid artery	SC	Spinal cord
IJV	Internal jugular vein	SCM	Sternocleidomastoid muscle
Ma	Masseter muscle		
MH	Mylohyoid muscle		
Mna	Mandible, angle		
MP	Medial pterygoid muscle		
Oph	Oropharynx		
PG	Parotid gland		

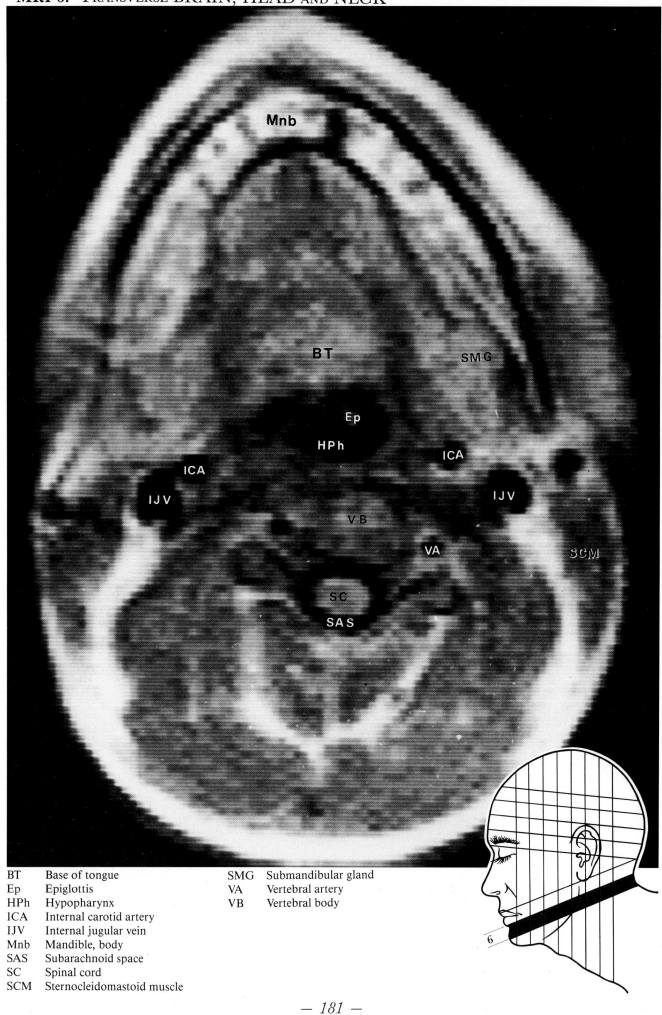

BT	Base of tongue	SMG	Submandibular gland
Ep	Epiglottis	VA	Vertebral artery
HPh	Hypopharynx	VB	Vertebral body
ICA	Internal carotid artery		
IJV	Internal jugular vein		
Mnb	Mandible, body		
SAS	Subarachnoid space		
SC	Spinal cord		
SCM	Sternocleidomastoid muscle		

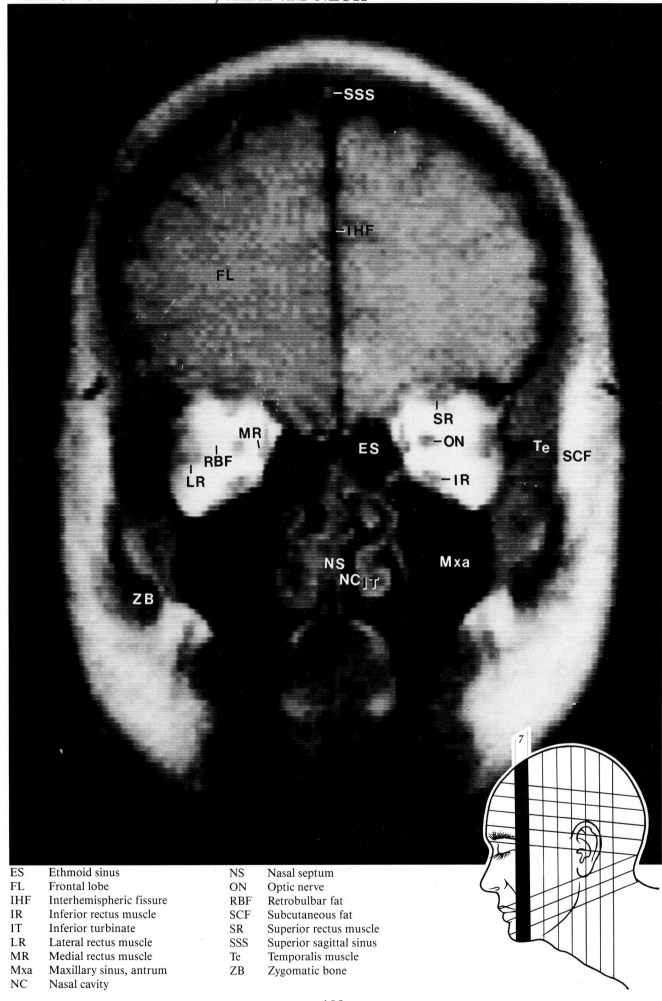

ES	Ethmoid sinus	NS	Nasal septum
FL	Frontal lobe	ON	Optic nerve
IHF	Interhemispheric fissure	RBF	Retrobulbar fat
IR	Inferior rectus muscle	SCF	Subcutaneous fat
IT	Inferior turbinate	SR	Superior rectus muscle
LR	Lateral rectus muscle	SSS	Superior sagittal sinus
MR	Medial rectus muscle	Te	Temporalis muscle
Mxa	Maxillary sinus, antrum	ZB	Zygomatic bone
NC	Nasal cavity		

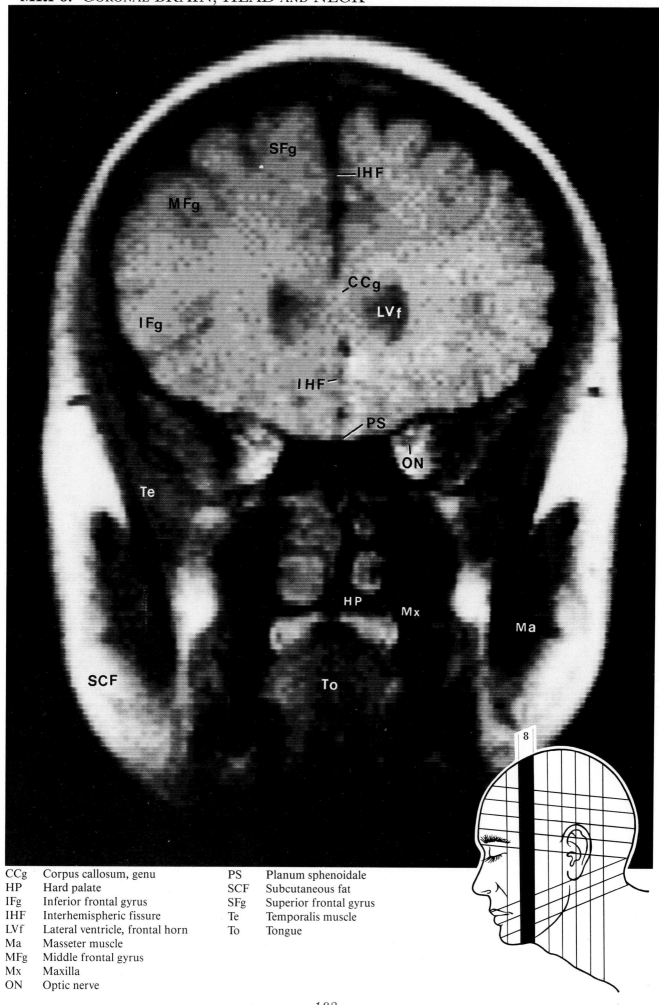

CCg	Corpus callosum, genu	PS	Planum sphenoidale
HP	Hard palate	SCF	Subcutaneous fat
IFg	Inferior frontal gyrus	SFg	Superior frontal gyrus
IHF	Interhemispheric fissure	Te	Temporalis muscle
LVf	Lateral ventricle, frontal horn	To	Tongue
Ma	Masseter muscle		
MFg	Middle frontal gyrus		
Mx	Maxilla		
ON	Optic nerve		

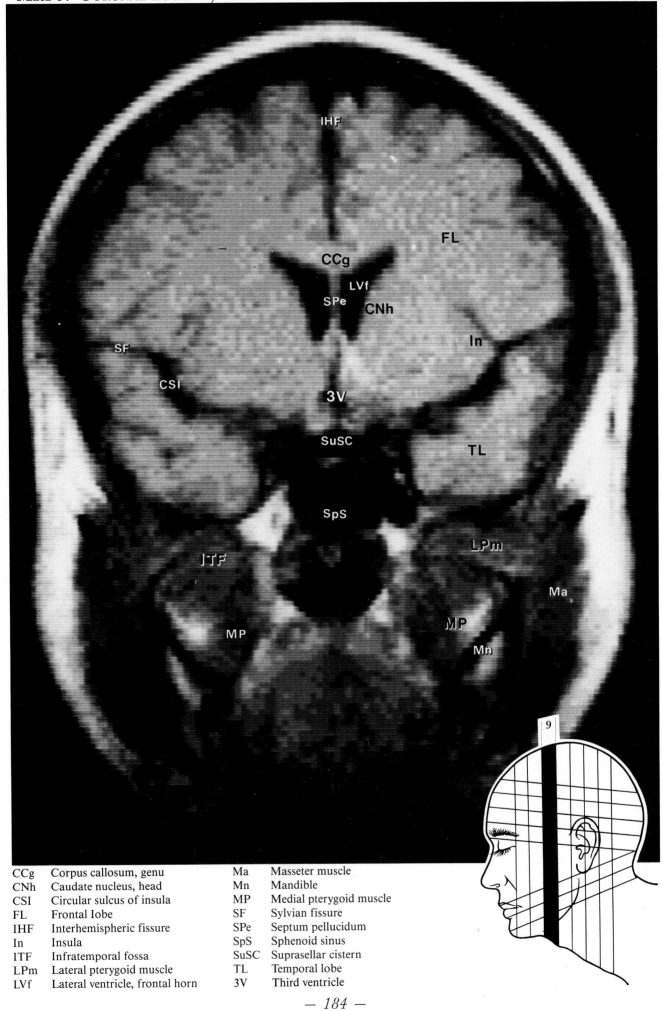

CCg	Corpus callosum, genu	Ma	Masseter muscle
CNh	Caudate nucleus, head	Mn	Mandible
CSI	Circular sulcus of insula	MP	Medial pterygoid muscle
FL	Frontal lobe	SF	Sylvian fissure
IHF	Interhemispheric fissure	SPe	Septum pellucidum
In	Insula	SpS	Sphenoid sinus
ITF	Infratemporal fossa	SuSC	Suprasellar cistern
LPm	Lateral pterygoid muscle	TL	Temporal lobe
LVf	Lateral ventricle, frontal horn	3V	Third ventricle

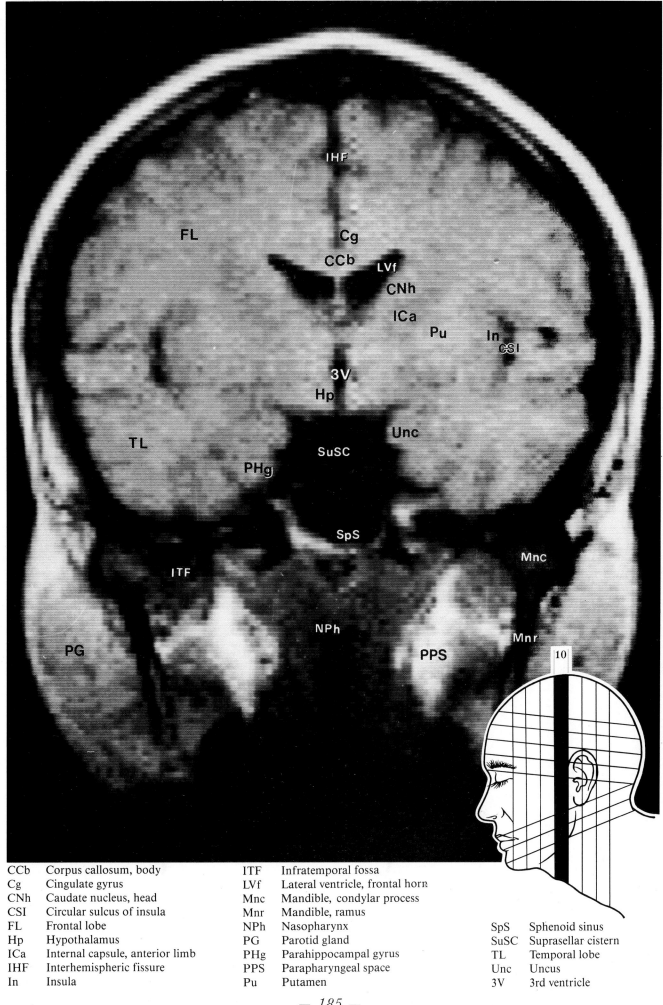

CCb	Corpus callosum, body	ITF	Infratemporal fossa
Cg	Cingulate gyrus	LVf	Lateral ventricle, frontal horn
CNh	Caudate nucleus, head	Mnc	Mandible, condylar process
CSl	Circular sulcus of insula	Mnr	Mandible, ramus
FL	Frontal lobe	NPh	Nasopharynx
Hp	Hypothalamus	PG	Parotid gland
ICa	Internal capsule, anterior limb	PHg	Parahippocampal gyrus
IHF	Interhemispheric fissure	PPS	Parapharyngeal space
In	Insula	Pu	Putamen

SpS	Sphenoid sinus
SuSC	Suprasellar cistern
TL	Temporal lobe
Unc	Uncus
3V	3rd ventricle

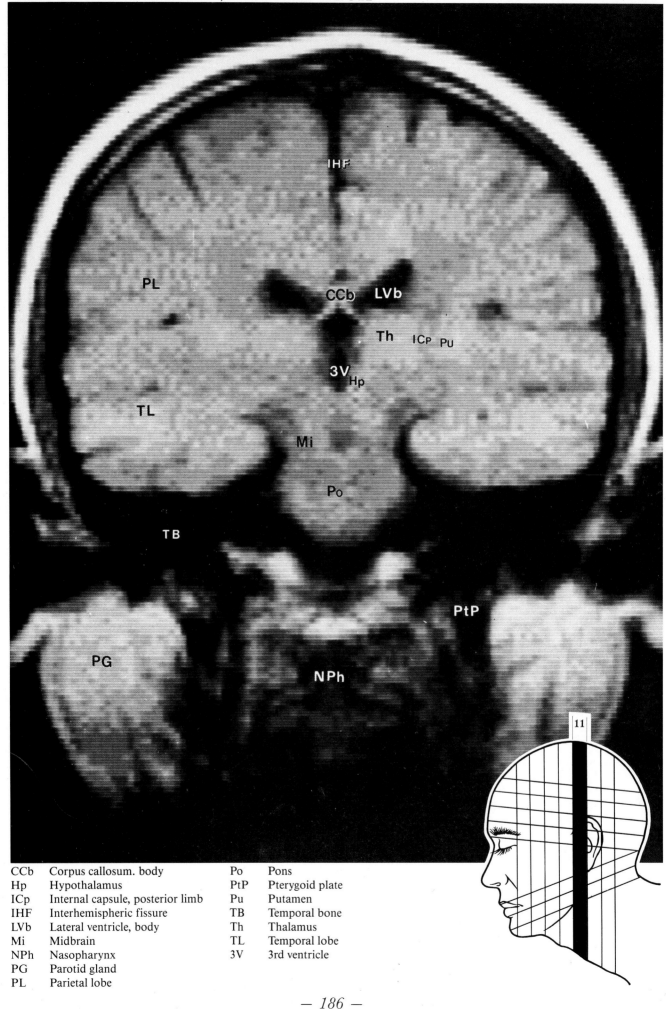

CCb	Corpus callosum. body	Po	Pons
Hp	Hypothalamus	PtP	Pterygoid plate
ICp	Internal capsule, posterior limb	Pu	Putamen
IHF	Interhemispheric fissure	TB	Temporal bone
LVb	Lateral ventricle, body	Th	Thalamus
Mi	Midbrain	TL	Temporal lobe
NPh	Nasopharynx	3V	3rd ventricle
PG	Parotid gland		
PL	Parietal lobe		

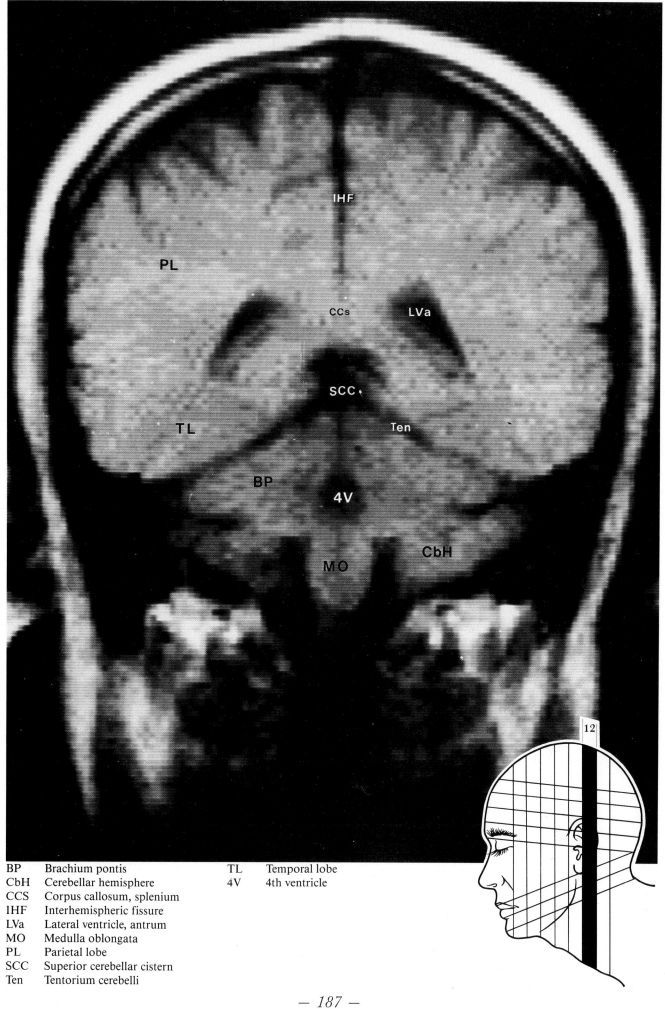

BP	Brachium pontis	TL	Temporal lobe
CbH	Cerebellar hemisphere	4V	4th ventricle
CCS	Corpus callosum, splenium		
IHF	Interhemispheric fissure		
LVa	Lateral ventricle, antrum		
MO	Medulla oblongata		
PL	Parietal lobe		
SCC	Superior cerebellar cistern		
Ten	Tentorium cerebelli		

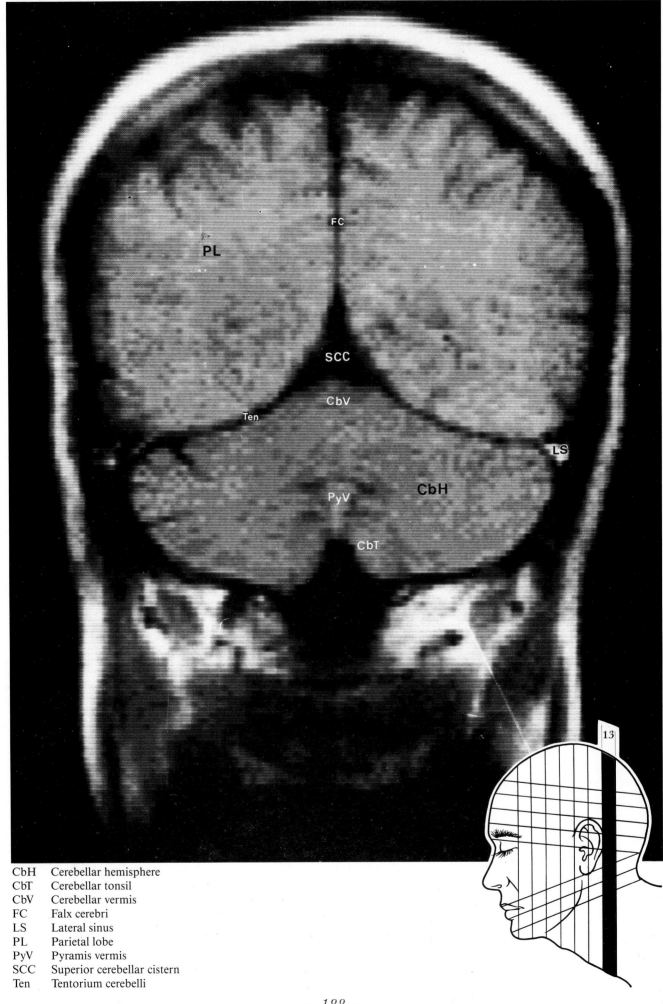

CbH   Cerebellar hemisphere
CbT   Cerebellar tonsil
CbV   Cerebellar vermis
FC   Falx cerebri
LS   Lateral sinus
PL   Parietal lobe
PyV   Pyramis vermis
SCC   Superior cerebellar cistern
Ten   Tentorium cerebelli

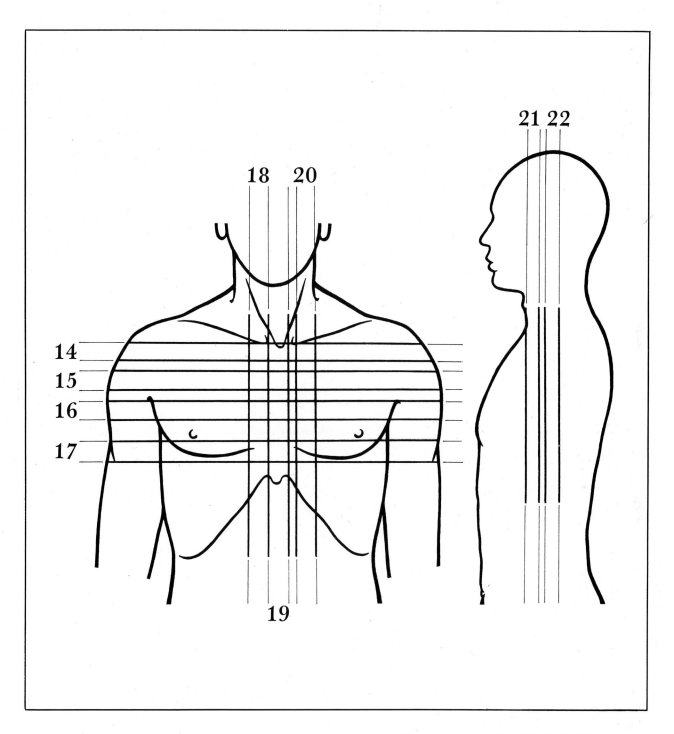

CHEST

Transverse MRI 14-17
Sagittal MRI 18-20
Coronal MRI 21-22

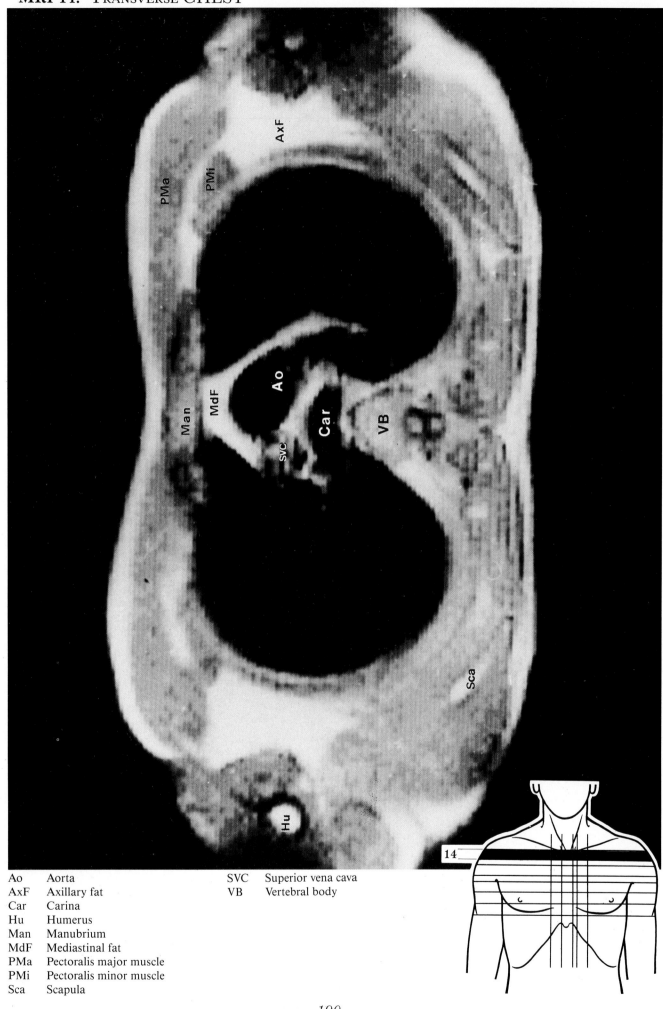

Ao	Aorta	SVC	Superior vena cava
AxF	Axillary fat	VB	Vertebral body
Car	Carina		
Hu	Humerus		
Man	Manubrium		
MdF	Mediastinal fat		
PMa	Pectoralis major muscle		
PMi	Pectoralis minor muscle		
Sca	Scapula		

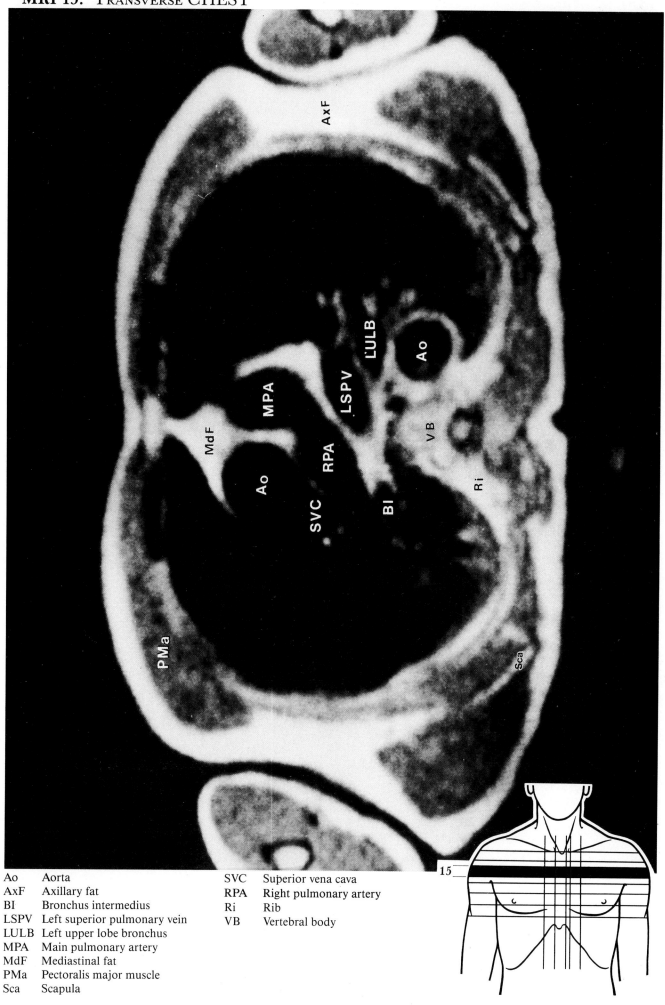

Ao	Aorta	SVC	Superior vena cava
AxF	Axillary fat	RPA	Right pulmonary artery
BI	Bronchus intermedius	Ri	Rib
LSPV	Left superior prulmonary vein	VB	Vertebral body
LULB	Left upper lobe bronchus		
MPA	Main pulmonary artery		
MdF	Mediastinal fat		
PMa	Pectoralis major muscle		
Sca	Scapula		

Ao	Aorta	SVC	Superior vena cava
LA	Left atrium of heart	Stn	Sternum
LAA	Left atrial appendage of heart	TMa	Teres major muscle
LSPV	Left superior pulmonary vein		
MdF	Mediastinal fat		
MPA	Main pulmonary artery		
RSPV	Right superior pulmonary vein		
RMLB	Right middle lobe bronchus		
SC	Spinal cord		

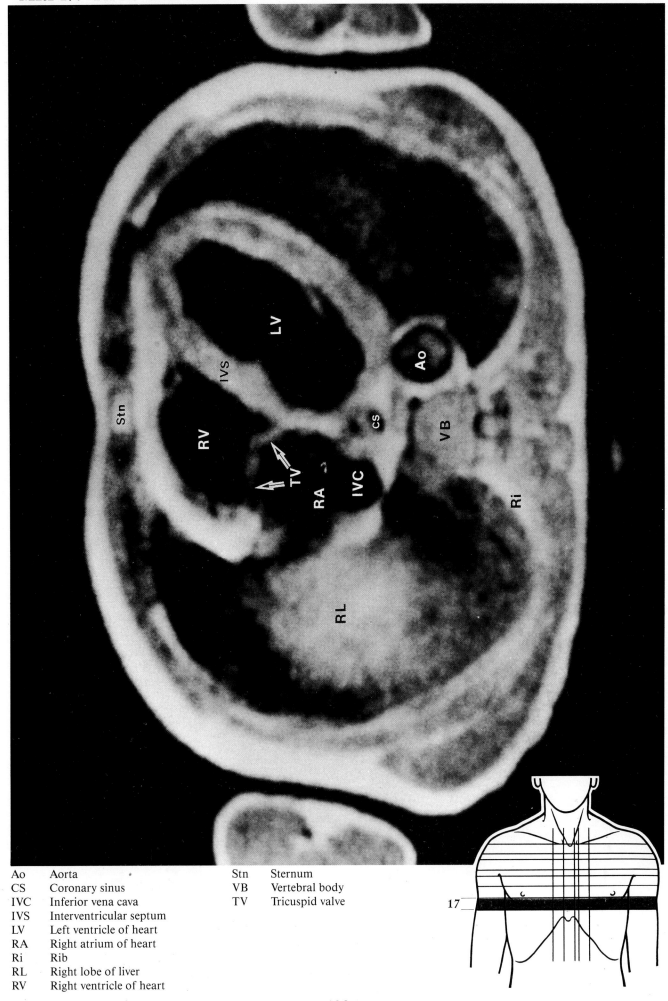

Ao	Aorta	Stn	Sternum
CS	Coronary sinus	VB	Vertebral body
IVC	Inferior vena cava	TV	Tricuspid valve
IVS	Interventricular septum		
LV	Left ventricle of heart		
RA	Right atrium of heart		
Ri	Rib		
RL	Right lobe of liver		
RV	Right ventricle of heart		

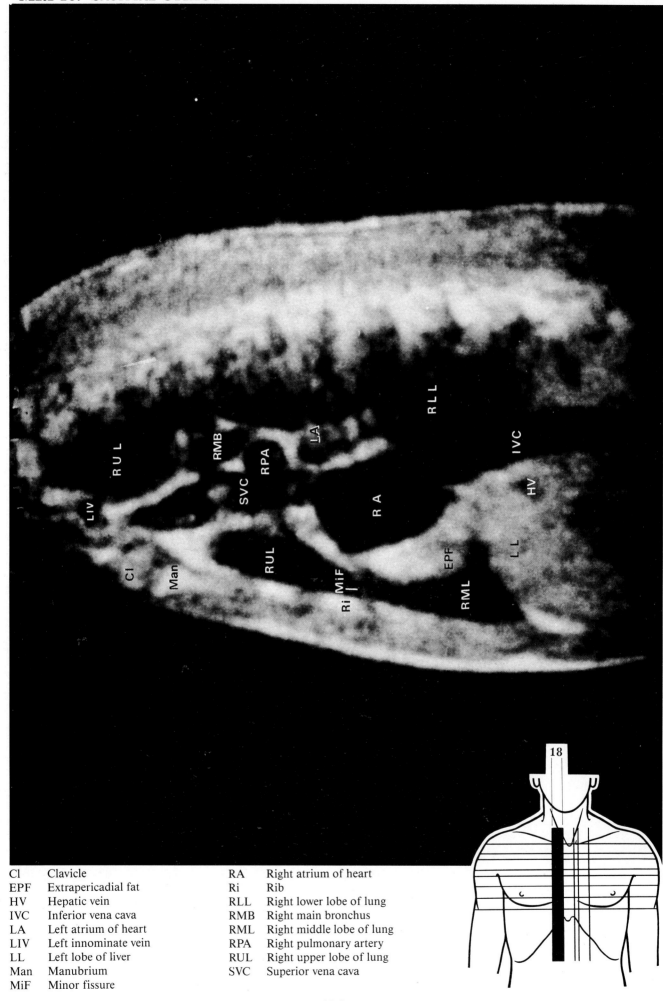

Cl	Clavicle	RA	Right atrium of heart
EPF	Extrapericadial fat	Ri	Rib
HV	Hepatic vein	RLL	Right lower lobe of lung
IVC	Inferior vena cava	RMB	Right main bronchus
LA	Left atrium of heart	RML	Right middle lobe of lung
LIV	Left innominate vein	RPA	Right pulmonary artery
LL	Left lobe of liver	RUL	Right upper lobe of lung
Man	Manubrium	SVC	Superior vena cava
MiF	Minor fissure		

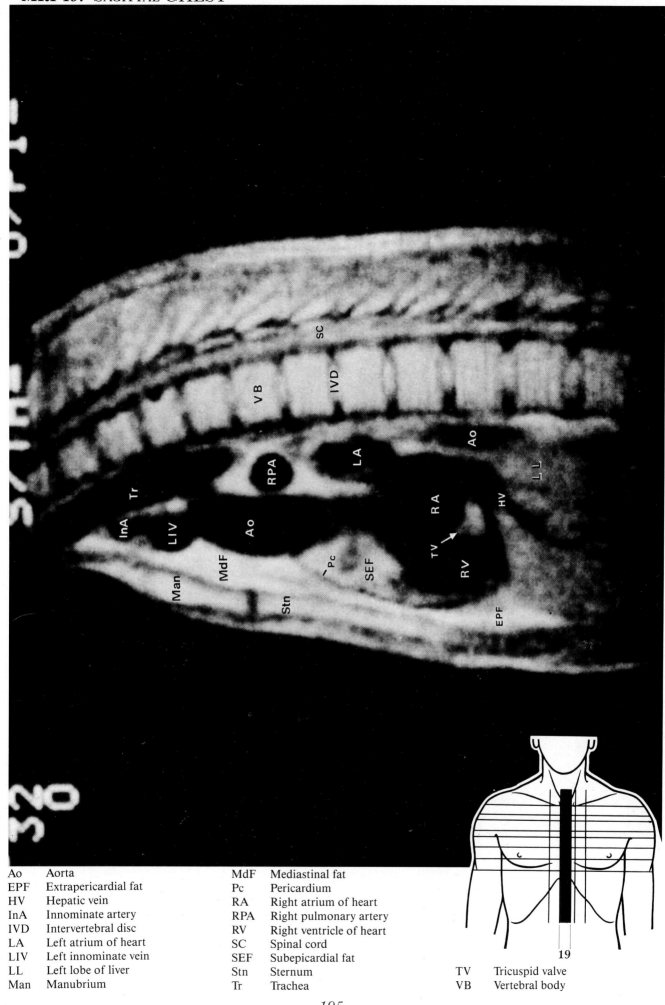

Ao	Aorta	MdF	Mediastinal fat		
EPF	Extrapericardial fat	Pc	Pericardium		
HV	Hepatic vein	RA	Right atrium of heart		
InA	Innominate artery	RPA	Right pulmonary artery		
IVD	Intervertebral disc	RV	Right ventricle of heart		
LA	Left atrium of heart	SC	Spinal cord		
LIV	Left innominate vein	SEF	Subepicardial fat		
LL	Left lobe of liver	Stn	Sternum	TV	Tricuspid valve
Man	Manubrium	Tr	Trachea	VB	Vertebral body

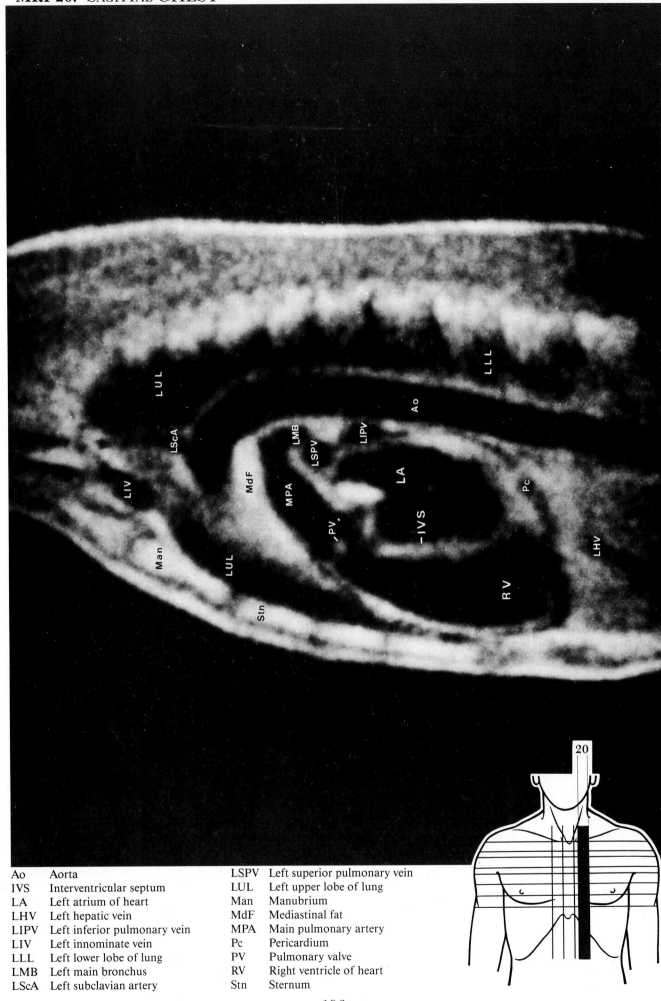

Ao	Aorta	LSPV	Left superior pulmonary vein
IVS	Interventricular septum	LUL	Left upper lobe of lung
LA	Left atrium of heart	Man	Manubrium
LHV	Left hepatic vein	MdF	Mediastinal fat
LIPV	Left inferior pulmonary vein	MPA	Main pulmonary artery
LIV	Left innominate vein	Pc	Pericardium
LLL	Left lower lobe of lung	PV	Pulmonary valve
LMB	Left main bronchus	RV	Right ventricle of heart
LScA	Left subclavian artery	Stn	Sternum

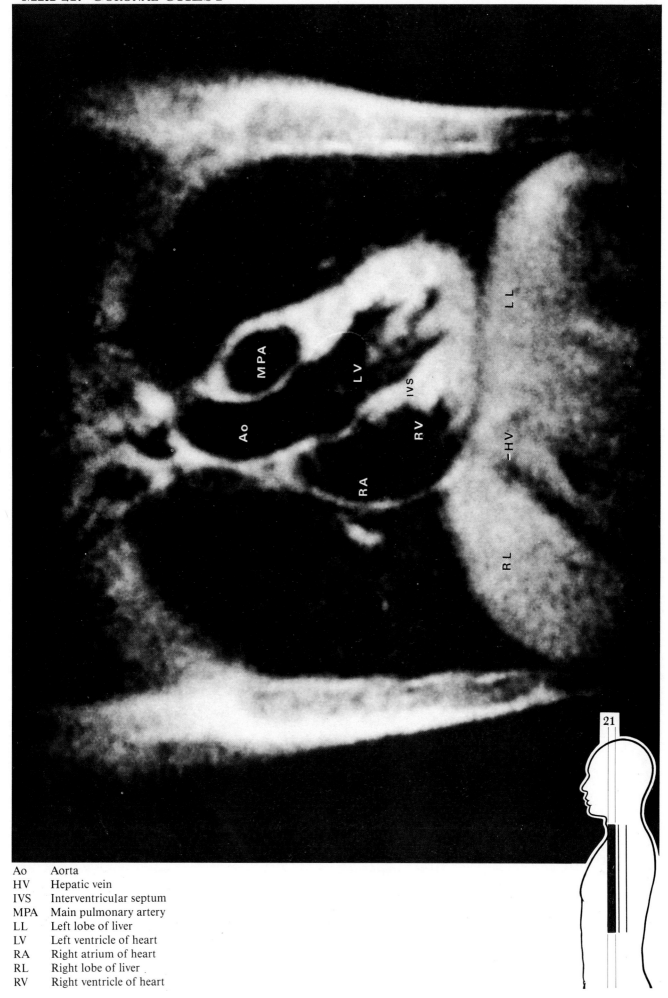

Ao	Aorta
HV	Hepatic vein
IVS	Interventricular septum
MPA	Main pulmonary artery
LL	Left lobe of liver
LV	Left ventricle of heart
RA	Right atrium of heart
RL	Right lobe of liver
RV	Right ventricle of heart

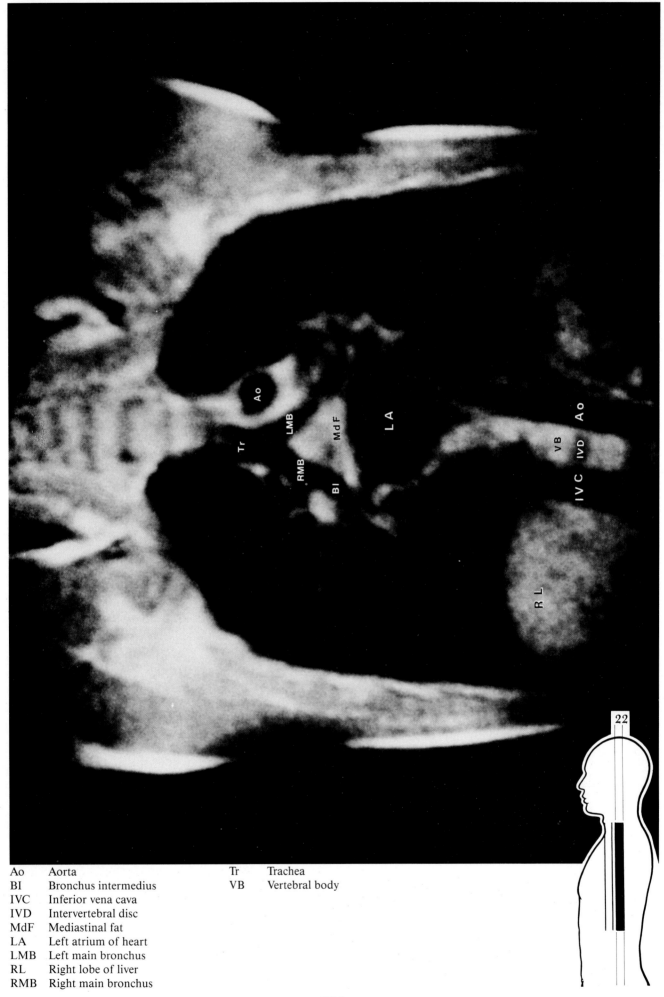

Ao	Aorta	Tr	Trachea
BI	Bronchus intermedius	VB	Vertebral body
IVC	Inferior vena cava		
IVD	Intervertebral disc		
MdF	Mediastinal fat		
LA	Left atrium of heart		
LMB	Left main bronchus		
RL	Right lobe of liver		
RMB	Right main bronchus		

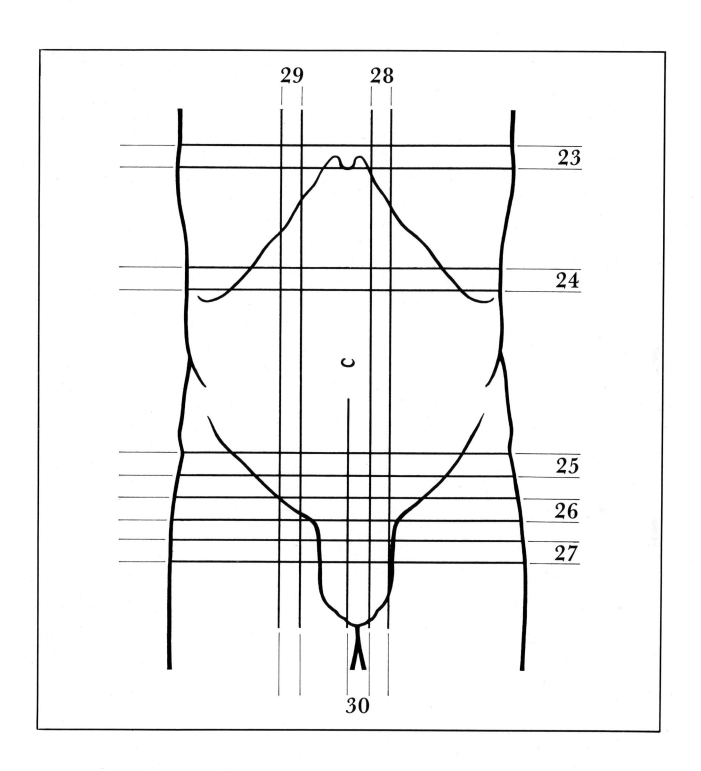

ABDOMEN AND PELVIS   Transverse MRI 23-27
                      Sagittal MRI 28-30

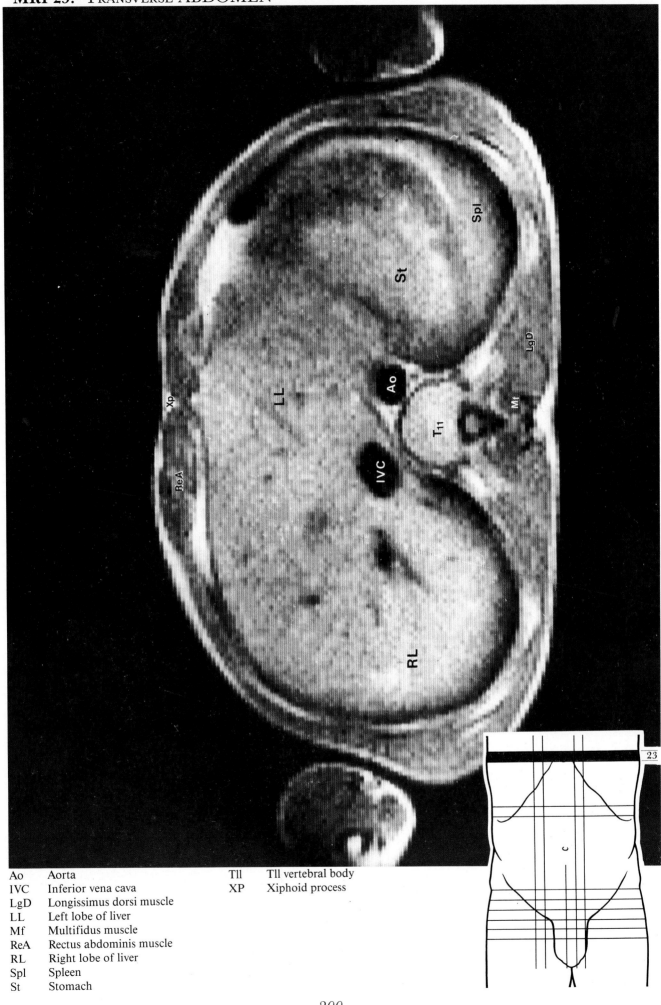

Ao	Aorta	Tll	Tll vertebral body
IVC	Inferior vena cava	XP	Xiphoid process
LgD	Longissimus dorsi muscle		
LL	Left lobe of liver		
Mf	Multifidus muscle		
ReA	Rectus abdominis muscle		
RL	Right lobe of liver		
Spl	Spleen		
St	Stomach		

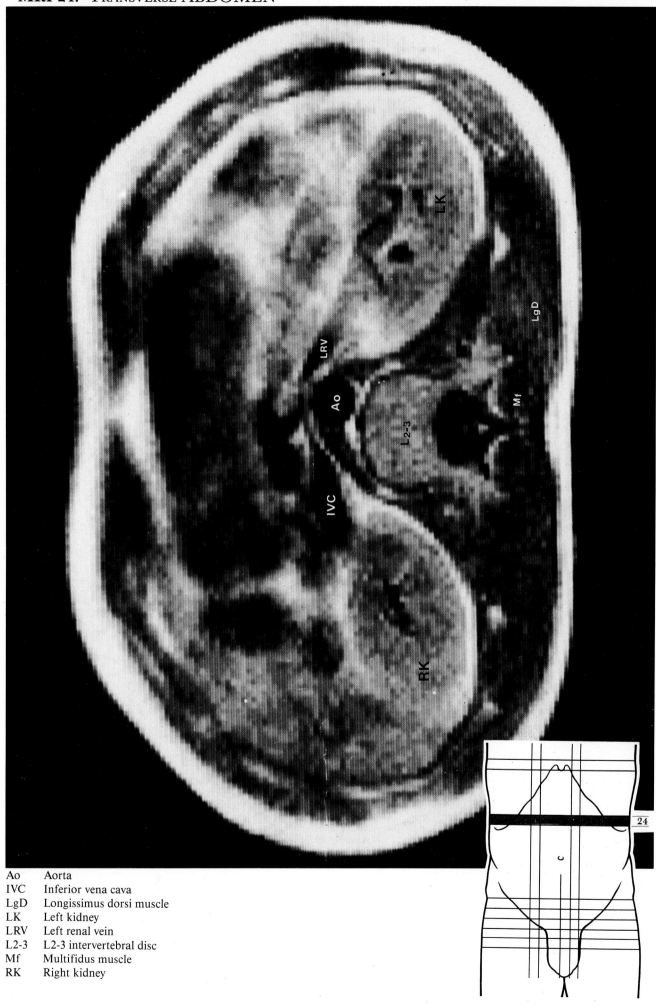

Ao   Aorta
IVC   Inferior vena cava
LgD   Longissimus dorsi muscle
LK   Left kidney
LRV   Left renal vein
L2-3   L2-3 intervertebral disc
Mf   Multifidus muscle
RK   Right kidney

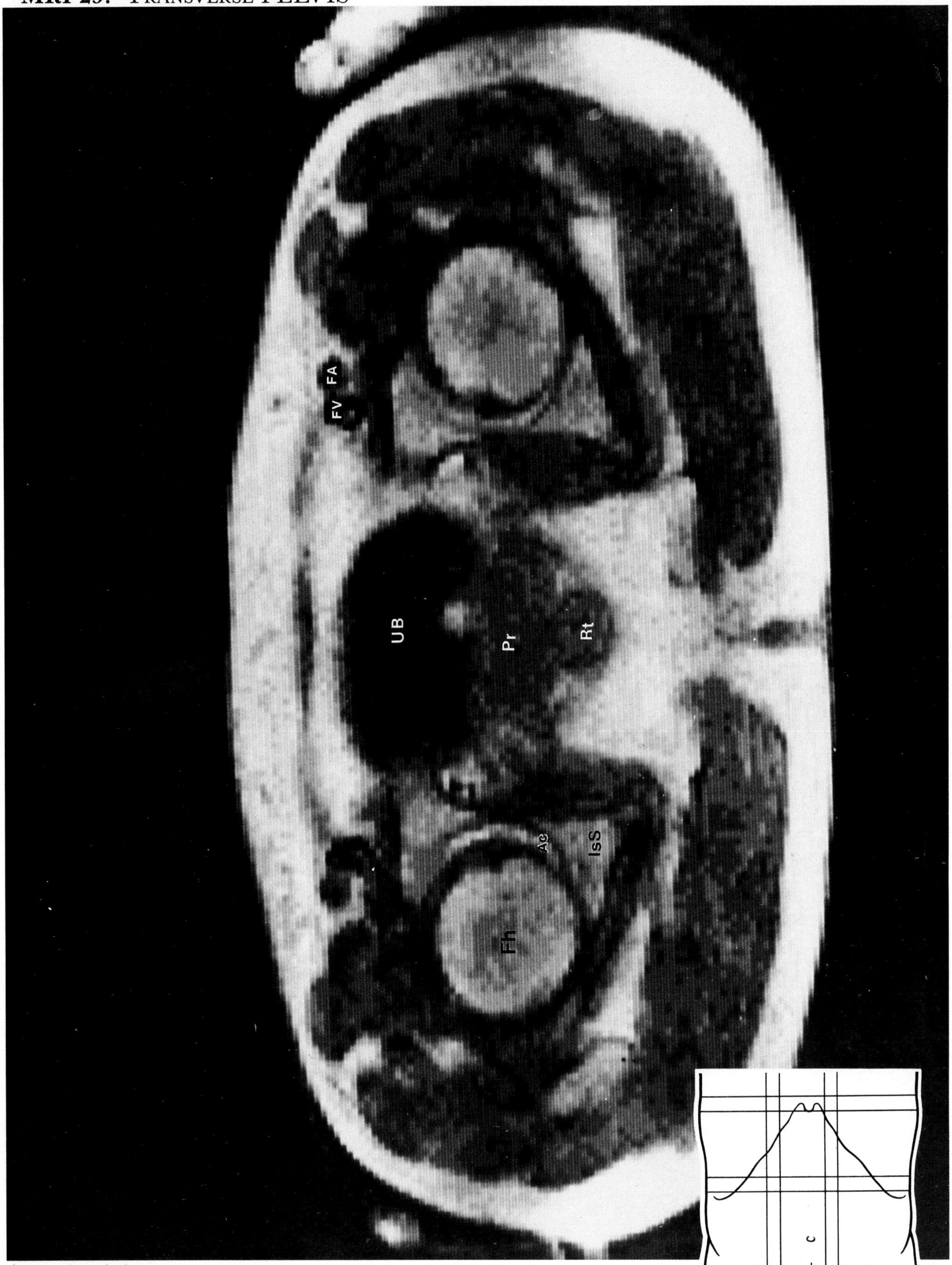

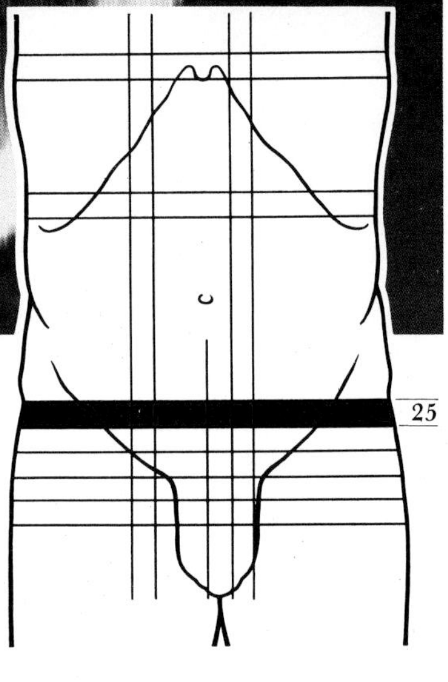

Ac   Acetabulum
FA   Femoral artery
Fh   Femur, head
FV   Femoral vein
IsS   Ischial spine
Pr   Prostate
Rt   Rectum
UB   Urinary bladder

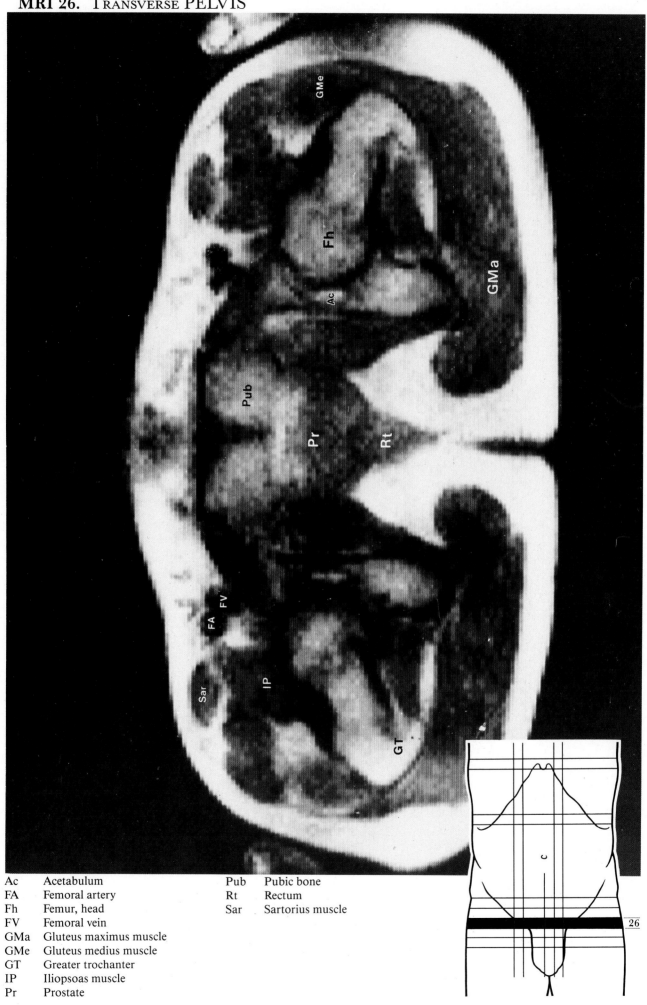

Ac   Acetabulum
FA   Femoral artery
Fh   Femur, head
FV   Femoral vein
GMa   Gluteus maximus muscle
GMe   Gluteus medius muscle
GT   Greater trochanter
IP   Iliopsoas muscle
Pr   Prostate

Pub   Pubic bone
Rt   Rectum
Sar   Sartorius muscle

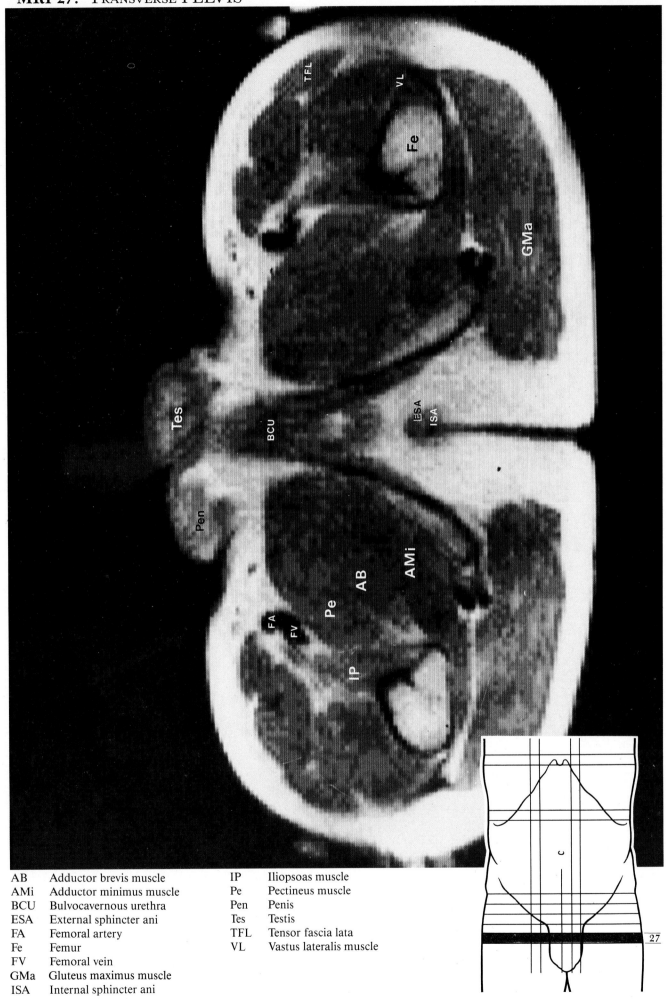

AB	Adductor brevis muscle	IP	Iliopsoas muscle
AMi	Adductor minimus muscle	Pe	Pectineus muscle
BCU	Bulvocavernous urethra	Pen	Penis
ESA	External sphincter ani	Tes	Testis
FA	Femoral artery	TFL	Tensor fascia lata
Fe	Femur	VL	Vastus lateralis muscle
FV	Femoral vein		
GMa	Gluteus maximus muscle		
ISA	Internal sphincter ani		

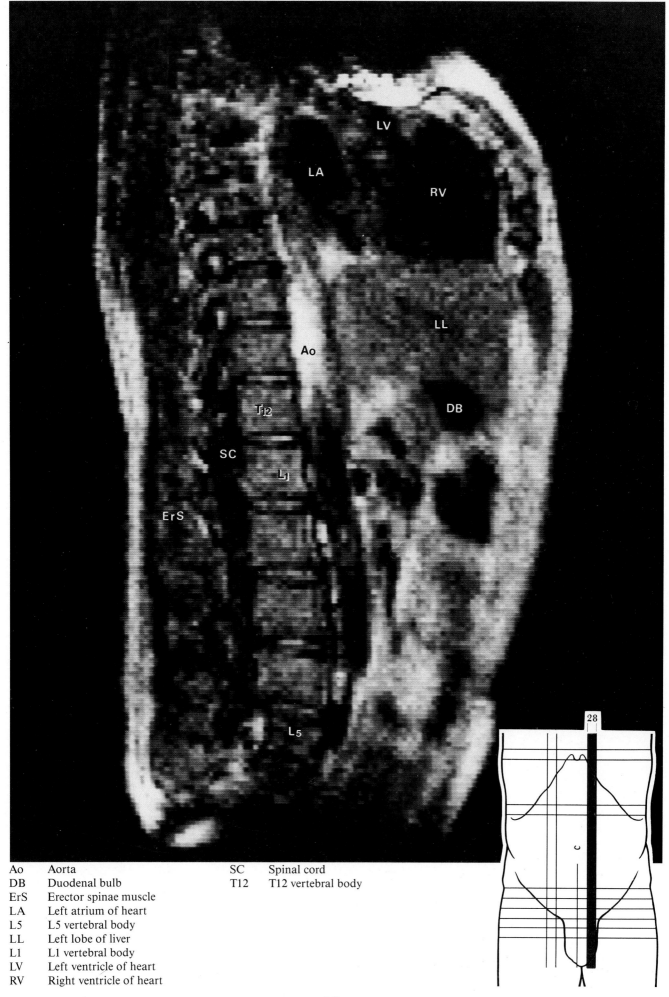

Ao	Aorta	SC	Spinal cord
DB	Duodenal bulb	T12	T12 vertebral body
ErS	Erector spinae muscle		
LA	Left atrium of heart		
L5	L5 vertebral body		
LL	Left lobe of liver		
L1	L1 vertebral body		
LV	Left ventricle of heart		
RV	Right ventricle of heart		

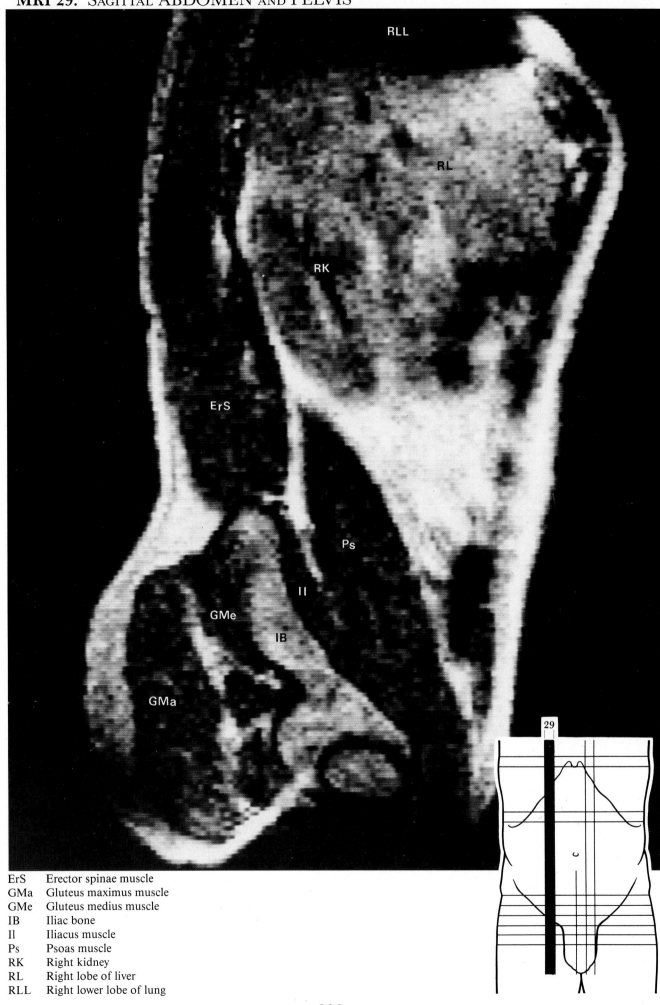

RLL

RL

RK

ErS

Ps

Il

GMe

IB

GMa

29

ErS   Erector spinae muscle
GMa   Gluteus maximus muscle
GMe   Gluteus medius muscle
IB   Iliac bone
Il   Iliacus muscle
Ps   Psoas muscle
RK   Right kidney
RL   Right lobe of liver
RLL   Right lower lobe of lung

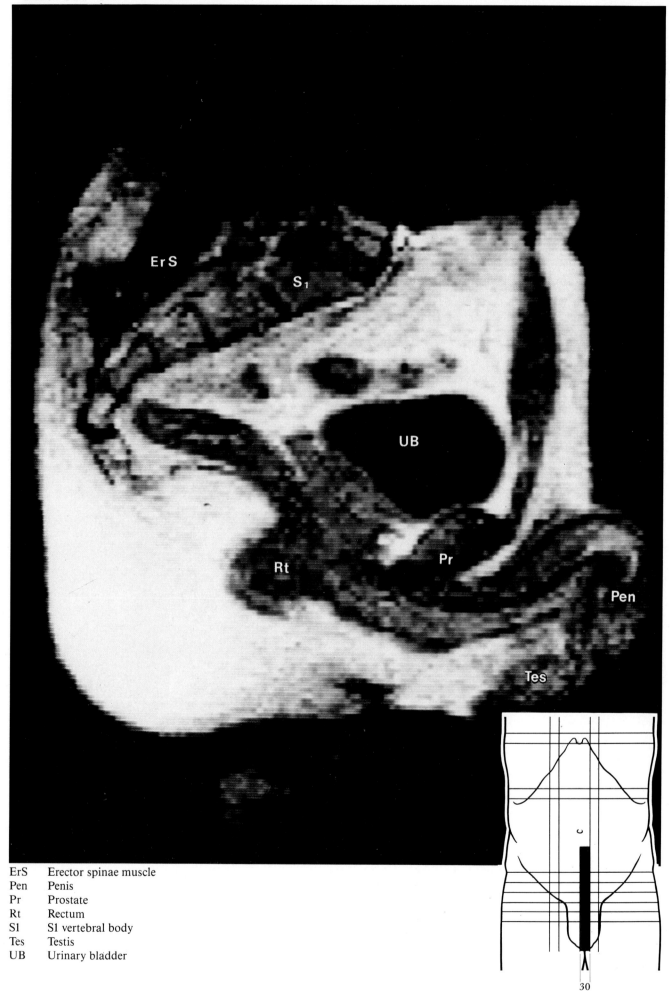

ErS   Erector spinae muscle
Pen   Penis
Pr   Prostate
Rt   Rectum
S1   S1 vertebral body
Tes   Testis
UB   Urinary bladder

INDEX

Left lower lobe bronchus	左下葉氣管枝	좌하엽기관지	LLLB	72~75, 90
Left lower lobe of lung	左肺下葉	좌폐하엽	LLL	68~83, 89~91, 96~103, 170, 171, 196
Left main bronchus	左主氣管枝	좌주기관지	LMB	70, 88, 89, 196, 198
Left portal vein	左門脈	좌문맥	LPV	89, 100, 165
Left pulmonary artery	左肺動脈	좌폐동맥	LPA	70, 74, 89, 90
Left renal artery	左腎動脈	좌신동맥	LRA	114~121, 168
Left renal vein	左腎靜脈	좌신정맥	LRV	114~121, 165~168, 201
Left subclavian artery	左鎖骨下動脈	좌쇄골하동맥	LScA	89~91, 196
Left subclavian vein	左鎖骨下靜脈	좌쇄골하정맥	LScV	91
Left superior pulmonary vein	左上肺靜脈	좌상폐정맥	LSPV	70~73, 89, 90, 191, 192, 196
Left upper lobe bronchus	左上葉氣管枝	좌상엽기관지	LULB	90, 191
Left upper lobe of lung	左肺上葉	좌폐상엽	LUL	66~73, 76~79, 89~91, 171, 196
Left ventricle of heart	左心室	좌심실	LV	74~83, 89~91, 171, 193, 197, 205
Lens	水晶體	수정체	Le	18
Lesser omentum	小網	소망	LOt	98, 102, 166, 167
Lesser sac	小囊	소낭	LSa	166, 167
Levator ani muscle	肛門擧筋	항문거근	LeA	146~155
Levator palati muscle	口蓋擧筋	구개거근	LP	22, 52
Levator scapularis muscle	肩胛擧筋	견갑거근	LeS	28~31, 34, 38~41
Ligament	**靭帶**	**인대**		
anterior sacroiliac ligament	前薦腸骨靭帶	전천장골인대	ASL	132
dentate ligament	齒狀靭帶	치상인대	DL	32
falciform ligament	鎌狀靭帶	겸상인대	FaL	110
fissure for ligamentum venosum	靜脈靭帶裂	정맥인대열	FLV	98, 102
gastrocolic ligament	胃大腸靭帶	위대장인대	GC	169
iliofemoral ligament	腸大腿骨靭帶	장대퇴골인대	IFL	144
inferior pulmonary ligament	下肺靭帶	하폐인대	IPL	80, 163
inguinal ligament	鼠蹊靭帶	서혜인대	Ing	162
interosseous sacroiliac ligament	骨間薦腸靭帶	골간천장인대	IoSL	132
ligamentum capitis femoris	大腿骨頭靭帶	대퇴골두인대	LCF	144
ligamentum flavum	黃色靭帶	황색인대	LF	132
ligamentum teres	圓形靭帶	원형인대	LT	165
medial umbilical ligament	內側臍靭帶	내측제인대	MlUL	134
median umbilical ligament	正中臍靭帶	정중제인대	MUL	134
nuchal ligament	項靭帶	항인대	NL	36, 40
sacrospinous ligament	薦骨脊椎間靭帶	천골척추간인대	SSL	144
vocal ligament	聲帶	성대	VoL	34
Ligamentum capitis femoris	大腿骨頭靭帶	대퇴골두인대	LCF	144
Ligamentum flavum	黃色靭帶	황색인대	LF	132
Ligamentum teres	圓形靭帶	원형인대	LT	165
Lingual gyrus	舌回	설회	Lg	56~59
Lingular segmental bronchus	舌小葉氣管枝	설소엽기관지	LSB	72
Liver	**肝**	**간**		
caudate lobe of liver	尾狀葉	미상엽	CL	88, 102, 166, 167
caudate process of caudate lobe	尾狀葉尾狀突起	미상엽미상돌기	CPC	104
left lobe of liver	肝左葉	간좌엽	LL	88~91, 96~103, 108, 165~168, 171, 194, 195, 197, 200, 205
papillary process of caudate lobe	尾狀葉乳頭突起	미상엽유두돌기	PPC	104
right lobe of liver	肝右葉	간우엽	RL	82, 86, 96~125, 162, 164, 197, 198, 200, 206
Long thoracic nerve	長胸神經	장흉신경	LTN	70
Longissimus dorsi muscle	背側最長筋	배측최장근	LgD	68~71, 74~83, 96~131, 200, 201
Longus colli and capitis muscle	頸長筋, 頭長筋	경장근및두장근	LCC	20~41
Lumbosacral trunk	腰薦骨神經幹	요천골신경간	LST	136
Lung	**肺**	**폐**		
anteromedial basal segment of lung	前內側基底肺分節	전내측기저폐분절	AMB	76
calcification of lung	肺石灰化	폐석회화	Cal	72
left lower lobe of lung	左肺下葉	좌폐하엽	LLL	68~83, 89~91, 96~103, 170, 171, 196
left upper lobe of lung	左肺上葉	좌폐상엽	LUL	66~73, 75~79, 89~91, 171, 196
posterolateral basal segment of lung	後外側基底肺分節	후외측기저폐분절	PBS	76
right lower lobe of lung	右肺下葉	우폐하엽	RLL	68~83, 86, 87, 96~99, 162, 165, 194, 206
right middle lobe of lung	右肺中葉	우폐중엽	RML	74~81, 162, 194
right upper lobe of lung	右肺上葉	우폐상엽	RUL	66~71, 86, 87, 194
Luschka joint	Luschka關節	루시카관절	LJ	32
Lymph node	淋巴節	림프절	LN	66~71, 86, 87, 89, 90
Main pulmonary artery	主肺動脈	주폐동맥	MPA	70~73, 89, 192, 196, 197
Major fissure	主裂溝	주열구	MaF	70~76, 80, 86, 90, 91, 162, 170, 171
Mammary gland	乳腺	유선	MG	91, 162, 171
Mandible	下顎骨	하악골	Mn	48~51, 184
Mandible, angle	下顎角	하악각	Mna	28, 180

Right renal artery	右腎動脈	우신동맥	RRA	114〜119, 164
Right renal vein	右腎靜脈	우신정맥	RRV	114〜117, 164
Right subclavian artery	右鎖骨下動脈	우쇄골하동맥	RScA	86
Right superior pulmonary vein	右上肺靜脈	우상폐정맥	RSPV	70〜73, 86, 192
Right upper lobe of lung	右肺上葉	우폐상엽	RUL	66〜71, 86, 87, 194
Right ventricle of heart	右心室	우심실	RV	74〜83, 88〜91, 167〜170, 193, 195〜197, 205
Root of penis	陰莖根	음경근	RPe	152〜155
Rosenmüller fossa	Rosenmüller 窩	로젠뮐러와	RF	22, 50〜53
S1 vertebral body	第一薦椎體	제 1 천추제	S1	207
Sacral nerve	薦骨神經	천골신경	SaN	132〜135
Sacroiliac joint	薦腸關節	천장관절	SIJ	132〜139
Sacrospinous ligament	薦骨脊椎間靭帶	천골척추간인대	SSL	144
Sacrum	薦骨	천골	Sac	132〜145, 163, 164, 168
Sartorius muscle	縫工筋	봉공근	Sar	140〜153, 156〜159, 203
Scapula	肩胛骨	견갑골	Sca	36〜41, 66〜69, 72〜77, 88, 91, 190, 191
Sciatic nerve	坐骨神經	좌골신경	SN	146, 150, 156
Sclera	鞏膜	공막	Scl	18
Scrotum	陰囊	음낭	Scr	156〜159
Second portion of duodenum	十二指腸第二部	십이지장제 2 부	D2	112〜117, 164
Seminal vesicle	精囊	정낭	SVe	146
Semispinalis capitis muscle	頭部半棘筋	두부반극근	SSC	26〜33, 60
Semispinalis cervicalis muscle	頸部半棘筋	경부반극근	SSCe	26〜33, 36〜41
Semispinalis dorsi muscle	背半棘筋	배반극근	SSD	66〜71, 74〜77, 82, 96〜105
Semitendinosus muscle	半腱狀筋	반건상근	Se	156
Septum pelludicum	透明中隔	투명중격	SPe	10〜13, 48〜53, 184
Serratus anterior muscle	前鋸筋	전거근	SeA	40, 67, 70〜79, 96〜103
Sigmoid colon	S狀結腸	S상결장	SiC	124〜137, 163, 165〜167, 169
Sigmoid sinus	S形洞	S형동	SS	17, 22, 54〜57
Soft palate	軟口蓋	연구개	SP	24〜27, 48〜55
Spermatic cord	精索	정색	SCo	138, 142〜155, 192
Sphenoid sinus	蝶形骨洞	접형골동	SpS	18, 48〜51, 184, 185
Spinal cord	脊髓	척수	SC	26〜39, 56〜61, 71, 80〜83, 96〜113, 180, 181, 195, 205
Spinal nerve ganglion	脊髓神經神經節	척수신경신경절	SNG	26, 34, 40
Spinalis cervicalis muscle	頸棘筋	경극근	SCe	36〜41
Spinalis muscle	脊椎棘筋	척추극근	Sp	96〜105, 169
Spleen	脾臟	비장	Spl	90, 91, 98〜113, 117, 170, 171, 200
Splenic artery	脾動脈	비동맥	SA	88, 106〜109, 167, 168
Splenic flexure of colon	脾結腸曲	비결장곡	SF	171
Splenic vein	脾靜脈	비정맥	SV	102〜105, 108〜111, 167〜169, 171
Splenius capitis muscle	頭部脊椎棘筋	두부척추극근	SpC	26〜33, 58, 60
Sternocleidomastoid muscle	胸鎖乳突筋	흉쇄유돌근	SCM	24〜41, 54〜57, 180, 181
Sternum	胸骨	흉골	Stn	70〜81, 88, 97, 192, 193, 195, 196
Stomach	胃臟	위장	St	89〜91, 96〜103, 109, 167, 169〜171, 200
Stomach, antrum	胃前庭部	위전정부	StA	102〜111, 166, 168
Straight sinus	直靜脈洞	직정맥동	StS	11, 33, 58〜61
Styloglossus muscle	莖突舌筋	경돌설근	SG	29, 33, 53
Stylohyoid muscle	莖突舌骨筋	경돌설골근	SH	30
Styloid muscle	莖突筋	경돌근	Sty	24
Styloid process	莖狀突起	경상돌기	Stp	22〜25
Subarachnoid space	蜘蛛膜下腔	지주막하강	SAS	26, 181
Subclavian artery	鎖骨下動脈	쇄골하동맥	ScA	40, 66
Subclavian vein	鎖骨下靜脈	쇄골하정맥	ScV	67
Subcutaneous fat	皮下脂肪	피하지방	SCF	176, 179, 182, 183
Subdural space	硬膜下腔	경막하강	SDS	4, 10〜13, 44
Subepicardial fat	心膜下脂肪	심막하지방	SEF	82, 91, 195
Sublingual gland	舌下腺	설하선	SLG	49
Submandibular gland	下顎下腺	하악하선	SMG	28〜33, 50〜53, 181
Subscapularis muscle	肩胛下筋	견갑하근	SbS	36〜41, 66〜71, 91
Substantia nigra	黑質	흑질	SNi	15
Sulcus	**溝**	**구**		
calcarine sulcus	鳥距溝	오거구	CaS	56〜61
central sulcus	中央溝	중앙구	CnS	6〜13, 52〜55
cingulate sulcus	帶狀回溝	대상회구	CiS	48, 54
circular sulcus of insula	島皮質圓形溝	도피질원형구	CSI	10〜15, 48〜53, 177, 184, 185
collateral sulcus	側部溝	측부구	CoS	56〜59
hippocampal sulcus	海馬溝	해마구	HS	52
intraparietal sulcus	頭頂內溝	두정내구	IPS	56
median sulcus	正中溝	정중구	MeS	55

Uncinate process of pancreas, head	膵臟鉤狀突起	췌장구상돌기	UnP	114～119
Uncus	鉤狀突起	구상돌기	Unc	14～17, 51, 178, 185
Ureter	尿管	뇨관	Ur	118～145, 164, 168
Urethra	尿道	뇨도	Ua	152～155
Urinary bladder	膀胱	방광	UB	136～149, 165～167, 202, 207
Uterus	子宮	자궁	Ut	165～167
Uvula	口蓋垂	구개수	Uv	26, 53, 59

Vagina	膣	질	Vg	165
Vallecula	谷	곡	Va	30～33
Vastus iniermedius muscle	中間廣筋	중간광근	VI	156～159
Vastus medialis muscle	內側廣筋	내측광근	VM	156～159
Vein	**靜脈**	**정맥**		
axillary vein	腋窩靜脈	액와정맥	AxV	68
azygos vein	奇靜脈	기정맥	Az	68～83, 87, 96～101, 104～113
common iliac vein	總腸骨靜脈	총장골정맥	CIV	130～135, 164～166, 168
diploic vein	板間靜脈	판간정맥	DV	4
external iliac vein	外腸骨靜脈	외장골정맥	EIV	136～145, 163
external jugular vein	外頸靜脈	외경정맥	EJV	30～41
femoral vein	大腿靜脈	대퇴정맥	FV	146～155, 158, 202～204
hemiazygos vein	半奇靜脈	반기정맥	HzV	72, 78～83, 88, 96～109, 112
hepatic vein	肝靜脈	간정맥	HV	87, 102, 165, 194, 195, 197
inferior gluteal vein	下臀靜脈	하둔정맥	IGV	142, 148～155
intercostal vein	肋間靜脈	늑간정맥	IcV	76, 80, 89, 91, 98～101
internal cerebral vein	內大腦靜脈	내대뇌정맥	ICV	10～13, 52～55
internal iliac vein	內腸骨靜脈	내장골정맥	IIV	136～141, 163, 168
internal jugular vein	內頸靜脈	내경정맥	IJV	20～41, 54, 180, 181
left gastric vein	左胃靜脈	좌위정맥	LGV	104～107
left hepatic vein	左肝靜脈	좌간정맥	LHV	88, 89, 96, 166～168, 196
left inferior pulmonary vein	左下肺靜脈	좌하폐정맥	LIPV	74～77, 196
left innominate vein	左無名靜脈	좌무명정맥	LIV	66, 87～90, 194, 196
left portal vein	左門脈	좌문맥	LPV	89, 100, 165
left renal vein	左腎靜脈	좌신정맥	LRV	114～121, 165～168, 201
left subclavian vein	左鎖骨下靜脈	좌쇄골하정맥	LScV	91
left superior pulmonary vein	左上肺靜脈	좌상폐정맥	LSPV	70～73, 89, 90, 191, 192, 196
middle cardiac vein	中心靜脈	중심정맥	MCV	88～90, 167
middle hepatic vein	中肝靜脈	중간정맥	MHV	96～99
obturator vein	閉鎖靜脈	폐쇄정맥	ObV	144～149
portal vein	門脈	문맥	PV	86, 87, 102～111, 164, 166
retromandibular vein	後下顎靜脈	후하악정맥	RMV	24～29, 180
right hepatic vein	右肝靜脈	우간정맥	RHV	86, 87, 96～101, 164
right inferior pulmonary vein	右下肺靜脈	우하폐정맥	RIPV	74～77
right innominate vein	右無名靜脈	우무명정맥	RIV	66, 86, 87
right portal vein	右門脈	우문맥	RPV	71, 102～107, 111, 163
right renal vein	右腎靜脈	우신정맥	RRV	114～117, 164
right superior pulmonary vein	右上肺靜脈	우상폐정맥	RSPV	70～73, 86, 192
splenic vein	脾靜脈	비정맥	SV	102～105, 108～111, 167～169, 171
subclavian vein	鎖骨下靜脈	쇄골하정맥	ScV	67
superior gluteal yein	上臀靜脈	상둔정맥	SGV	136～143
superior mesenteric vein	上腸間膜靜脈	상장간막정맥	SMV	112～127, 167～168
superior ophthalmic vein	上眼靜脈	상안정맥	SOV	44～49
vein of Galen	Galen 靜脈	갈렌정맥	VG	57
vertebral vein	椎骨靜脈	추골정맥	VV	57
Ventricle	**腦室**	**뇌실**		
3rd ventricle	第三腦室	제삼뇌실	3V	12～15, 52, 177, 184～186
4th ventricle	第四腦室	제사뇌실	4V	16, 56, 179, 187
choroid plexus of lateral ventricle	側腦室脈絡叢	측뇌실맥락총	ChP	10, 54～57
lateral ventricle, antrum	側腦室洞	측뇌실동	LVa	10, 55, 187
lateral ventricle, body	側腦室體部	측뇌실체부	LVb	6～11, 52, 176, 186
lateral ventricle, frontal horn	側腦室前頭角	측뇌실전두각	LVf	10～15, 48～53, 177, 183～186
lateral ventricle, occipital horn	側腦室後頭角	측뇌실후두각	LVo	8～11, 56, 177
lateral ventricle, temporal horn	側腦室側頭角	측뇌실측두각	LVt	14～17, 52, 178
Vertebra	**脊椎**	**척추**		
C3 vertebral body	第三頸椎體	제 3 경추체	C3	28, 56
C3-4 intervertebral disc	第三，四頸椎間板	제 3 - 4 경추간판	C3-4	30
C4 vertebral body	第四頸椎體	제 4 경추체	C4	32
C4-5 intervertebral disc	第四，五頸椎間板	제 4 - 5 경추간판	C4-5	34
C5 vertebral body	第五頸椎體	제 5 경추체	C5	36
C6 vertebral body	第六頸椎體	제 6 경추체	C6	38～41